Life
Illuminated

Selected Papers from Cold Spring Harbor
Volume 2, 1972–1994

Life
Illuminated

Selected Papers from Cold Spring Harbor
Volume 2, 1972–1994

EDITED BY

Jan A. Witkowski

Cold Spring Harbor Laboratory, New York

Alexander Gann

Cold Spring Harbor Laboratory, New York

Joseph F. Sambrook

*Peter MacCallum Cancer Institute,
East Melbourne, Australia*

COLD SPRING HARBOR LABORATORY PRESS
Cold Spring Harbor, New York · www.cshlpress.com

Life Illuminated
Selected Papers from Cold Spring Harbor, Volume 2, 1972–1994

Publisher	John Inglis
Acquisition Editor	John Inglis
Book Development, Marketing, and Sales Director	Jan Argentine
Project Coordinator	Maryliz Dickerson
Permissions Coordinators	Carol Brown, Maria Fairchild
Production Editor	Mala Mazzullo
Copy Editor	Dotty Brown
Desktop Editor	Lauren Heller
Production Manager	Denise Weiss
Interior Designer	Denise Weiss
Cover Designer	Ed Atkeson

Front cover: First row: Ahmad Bukhari; Winship Herr; Birgit Zipser; Klaus Weber.
Second row: Mitch Goldfarb and Daniel Birnbaum; Mike Wigler and Scott Powers; Jim Hicks, Amar Klar, and Jeff Strathern; and Guenter Albrecht-Buehler.

Back cover: First row: Blackford; James; McClintock; Jones
Second row: Beckman; Delbrück; Grace; Demerec

Library of Congress Cataloging-in-Publication Data

Life illuminated : selected papers from Cold Spring Harbor : volume 2, 1972-1994 / edited by Jan A. Witkowski, Alexander Gann, Joseph F. Sambrook.
    p. ; cm.
  Vol. 1 has title: Illuminating life : selected papers from Cold Spring Harbor Laboratory, 1903-1969.
  Contains reprints of articles originally published in various journals from 1972-1994.
  Includes bibliographical references and indexes.
  ISBN 978-0-87969-804-1 (cloth : alk. paper)
  1. Molecular biology. 2. Molecular genetics. I. Witkowski, J. A. (Jan Anthony), 1947- II. Gann, Alexander. III. Sambrook, Joseph. IV. Cold Spring Harbor Laboratory. V. Illuminating life.
  [DNLM: 1. Cold Spring Harbor Laboratory. 2. Biomedical Research–Collected Works. 3. Molecular Biology–Collected Works. 4. Clinical Laboratory Techniques–Collected Works. 5. History, 20th Century. 6. Laboratories–history. QU 450 L722 2008]
  QH506.L543 2008
  572.8–dc22

                                                      2008007286

10 9 8 7 6 5 4 3 2 1

All Cold Spring Harbor Laboratory Press publications may be ordered directly from Cold Spring Harbor Laboratory Press, 500 Sunnyside Blvd., Woodbury, New York 11797-2924. Phone: 1-800-843-4388 in Continental U.S. and Canada. All other locations: (516) 422-4100. FAX: (516) 422-4097. E-mail: cshpress@cshl.edu. For a complete catalog of all Cold Spring Harbor Laboratory Press publications, visit our World Wide Web site http://www.cshlpress.com/.

CONTENTS

Preface xi

    Jan A. Witkowski, Alexander Gann, and Joseph F. Sambrook

Acknowledgments xv

GENETICS

Gerald R. Fink 1

    Introduction: Genes Are Interesting to Geneticists

Rasika Harshey 7

    Ljungquist E. and Bukhari A.I. 1977. State of prophage Mu DNA
    upon induction. *Proc. Natl. Acad. Sci.* **74:** 3143–3147.

Bukhari Memoriam 12

    Taylor A.L. and Szybalski W. 1984. In memoriam: Ahmad I.
    Bukhari, 1943–1983. *Gene* **27:** iii–iv.

James B. Hicks 13

    Strathern J.N, Klar A.J.S., Hicks J.B., Abraham J.A., Ivy J.M.,
    Nasmyth K.A., and McGill C. 1982. Homothallic switching of yeast
    mating type cassettes is initiated by a double-stranded cut in the
    MAT locus. *Cell* **31:** 183–192.

Robert Martienssen 19

    Sundaresan V., Springer P., Volpe T., Haward S., Jones J.D.G., Dean
    C., Ma H., and Martienssen R. 1995. Patterns of gene action in plant
    development revealed by enhancer trap and gene trap transposable
    elements. *Genes Dev.* **9:** 1797–1810.

David Beach 25

    Draetta G., Luca F., Westendorf J., Brizuela L., Ruderman J., and Beach
    D. 1989. cdc2 protein kinase is complexed with both cyclin A and B:
    Evidence for proteolytic inactivation of MPF. *Cell* **56:** 829–838.

DNA

Bruce Alberts   31

Introduction: Why the Revolution That Jim Started Continues

Joseph F. Sambrook   37

Watson J.D. 1972. Origin of concatemeric T7 DNA. *Nat. New Biol.* **239:** 197–201.

Bruce Stillman   41

Prelich G., Kostura M., Marshak D.R., Mathews M.B., and Stillman B. 1987. The cell-cycle regulated proliferating cell nuclear antigen is required for SV40 DNA replication *in vitro. Nature* **326:** 471–475.

Bruce Stillman   47

Bell S.P. and Stillman B. 1992. ATP-dependent recognition of eukaryotic origins of DNA replication by a multiprotein complex. *Nature* **357:** 128–134.

Richard J. Roberts   51

Klimašauskas S., Kumar S., Roberts R.J., and Cheng X. 1994. HhaI methyltransferase flips its target base out of the DNA helix. *Cell* **76:** 357–369.

CELL BIOLOGY

David L. Spector   55

Introduction: From Architecture to Functional Analysis

Klaus Weber   61

Lazarides E. and Weber K. 1974. Actin antibody: The specific visualization of actin filaments in non-muscle cells. *Proc. Natl. Acad. Sci.* **71:** 2268–2272.

Guenter Albrecht-Buehler   65

Albrecht-Buehler G. 1977. Daughter 3T3 cells. Are they mirror images of each other? *J. Cell Biol.* **72:** 595–603.

Dafna Bar-Sagi and James R. Feramisco   69

Bar-Sagi D. and Feramisco J.R. 1985. Microinjection of the *ras* oncogene protein into PC12 cells induces morphological differentiation. *Cell* **42:** 841–848.

Carol W. Greider   75

Harley C.B., Futcher A.B., and Greider C.W. 1990. Telomeres shorten during ageing of human fibroblasts. *Nature* **345:** 458–460.

David L. Spector 83

Spector D.L. 1990. Higher order nuclear organization: Three-dimensional distribution of small nuclear ribonucleoprotein particles. *Proc. Natl. Acad. Sci.* **87**: 147–151.

TUMOR VIRUSES

Joseph F. Sambrook 89

Introduction: The Early Days of Tumor Virus Research at Cold Spring Harbor

Joseph F. Sambrook 95

Sambrook J., Sharp P.A., and Keller W. 1972. Transcription of simian virus 40. I. Separation of the strands of SV40 DNA and hybridization of the separated strands to RNA extracted from lytically infected and transformed cells. *J. Mol. Biol.* **70**: 57–71.

Terri Grodzicker 99

Grodzicker T., Williams J., Sharp P., and Sambrook J. 1974. Physical mapping of temperature-sensitive mutations of adenoviruses. *Cold Spring Harbor Symp. Quant. Biol.* **39(Pt 1)**: 439–446.

Louise T. Chow 105

Chow L.T., Gelinas R.E., Broker T.R., and Roberts R.J. 1977. An amazing sequence arrangement at the 5' ends of adenovirus 2 messenger RNA. *Cell* **12**: 1–8.

Michael Botchan 113

Botchan M., Topp W., and Sambrook J. 1976. The arrangement of simian virus 40 sequences in the DNA of transformed cells. *Cell* **9**: 269–287.

Robert Tjian 119

Tjian R. 1978. The binding site on SV40 DNA for a T antigen-related protein. *Cell* **13**: 165–179.

Joseph F. Sambrook 125

Gluzman Y. 1981. SV40-transformed simian cells support the replication of early SV40 mutants. *Cell* **23**: 175–182.

Carl S. Thummel 129

Thummel C., Tjian R., and Grodzicker T. 1981. Expression of SV40 T antigen under control of adenovirus promoters. *Cell* **23**: 825–836.

Winship Herr   133

    Herr W. and Clarke J. 1986. The SV40 enhancer is composed of multiple functional
    elements that can compensate for one another. *Cell* **45:** 461–470.

NEUROSCIENCE

Eric Kandel   139

    Introduction: Neuroscience at CSH Laboratory: Meetings, Courses, and Research

Ron McKay   145

    Introduction: Getting the Point at Cold Spring Harbor

Birgit Zipser   151

    Zipser B. and McKay R. 1981. Monoclonal antibodies distinguish identifiable neu-
    rones in the leech. *Nature* **289:** 549–554.

Ron McKay   157

    Hockfield S. and McKay R.D.G. 1985. Identification of major cell classes in the
    developing mammalian nervous system. *J. Neurosci.* **5:** 3310–3328.

Tim Tully   163

    Yin J.C.P., Wallach J.S., Del Vecchio M., Wilder E.L., Zhou H., Quinn W.G., and Tully
    T. 1994. Induction of a dominant negative CREB transgene specifically blocks long-
    term memory in Drosophila. *Cell* **79:** 49–58.

CANCER

Arnold J. Levine   169

    Introduction: The Road to Understanding the Origins of Cancer in Humans

Mitchell Goldfarb   175

    Goldfarb M., Shimizu K., Perucho M., and Wigler M. 1982. Isolation and prelimi-
    nary characterization of a human transforming gene from T24 bladder carcinoma
    cells. *Nature* **296:** 404–409.

Earl Ruley   181

    Ruley H.E. 1983. Adenovirus early region 1A enables viral and cellular transform-
    ing genes to transform primary cells in culture. *Nature* **304:** 602–606.

Scott Powers   187

    Kataoka T., Powers S., Cameron S., Fasano O., Goldfarb M., Broach J., and Wigler M.
    1985. Functional homology of mammalian and yeast *RAS* genes. *Cell* **40:** 19–26.

Douglas Hanahan   193

> Hanahan D. 1985. Heritable formation of pancreatic β-cell tumours in transgenic mice expressing recombinant insulin/simian virus 40 oncogenes. *Nature* **315:** 115–122.

B. Robert Franza, Jr.   199

> Franza B.R. Jr., Rauscher F.J. 3rd, Josephs S.F., and Curran T. 1988. The Fos complex and Fos-related antigens recognize sequence elements that contain AP-1 binding sites. *Science* **239:** 1150–1153.

Ed Harlow   203

> Whyte P., Buchkovich K.J., Horowitz J.M, Friend S.H., Raybuck M., Weinberg R.A., and Harlow E. 1988. Association between an oncogene and an anti-oncogene: the adenovirus E1A proteins bind to the retinoblastoma gene product. *Nature* **334:** 124–129.

TECHNIQUES

Angela N.H. Creager   209

> Introduction: Early Technological Breakthroughs at Cold Spring Harbor

Richard J. Roberts   215

> Roberts R.J. 1978. Restriction and modification enzymes and their recognition sequences. *Gene* **4:** 183–194.

James I. Garrels   219

> Garrels J.I. 1989. The QUEST system for quantitative analysis of two-dimensional gels. *J. Biol. Chem.* **264:** 5269–5282.

Joseph F. Sambrook   223

> Sharp P.A., Sugden B., and Sambrook J. 1973. Detection of two restriction endonuclease activities in *Haemophilus parainfluenzae* using analytical agarose–ethidium bromide electrophoresis. *Biochemistry* **12:** 3055–3063.

Tom Maniatis   227

> Maniatis T., Kee S.G., Efstratiadis A., and Kafatos F.C. 1976. Amplification and characterization of a β-globin gene synthesized in vitro. *Cell* **8:** 163–182.

Author Index   235

Index   237

Cold Spring Harbor Laboratory, early 1970s. *(Courtesy of Cold Spring Harbor Laboratory Archives.)*

Cold Spring Harbor Laboratory, early 1990s. *(Courtesy of Cold Spring Harbor Laboratory Archives.)*

PREFACE

T HE PREVIOUS COLLECTION OF NOTABLE PAPERS from Cold Spring Harbor covered a period of 66 years, during which there were no fewer than six directors of the various institutes on the south west shore of the harbor—The Biological Laboratory, founded in 1890; Station for Experimental Evolution (later the Department of Genetics) of the Carnegie Institution of Washington (1904); and, finally, Cold Spring Harbor Laboratory created from a fusion of the earlier institutes, in 1962. The papers in this volume cover the directorship of a single director, James D. Watson.

Watson had had a long association with the Laboratory. He had first visited the harbor campus in 1948 when a graduate student with Salvador Luria, at the University of Indiana. Luria with Max Delbrück had by then spent several summers at Cold Spring Harbor, working on the genetics of bacteriophage, and in 1945 had established the Phage Course there. The course was intended not only to provide training in the new techniques of phage genetics, but also to promote phage genetics as the new, hottest area of genetics research. It is not clear that Watson learned much when he took the course that first summer, but his visit had two lasting effects. First, as a protégé of Luria, and then also of Delbrück, Watson had a ready introduction to the elite of the world of bacterial and phage genetics, a circle of friends and colleagues united in a no-holds-barred approach to doing science. Second, Watson fell in love with Cold Spring Harbor, although the buildings were dilapidated and the grounds in disarray. He returned to Cold Spring Harbor every year to attend meetings and courses or to meet colleagues. Most notably, he gave the first public presentation of the DNA double helix in 1953 at the 18th Cold Spring Harbor Symposium on Quantitative Biology, that year on the topic of Viruses.

Watson joined the Harvard faculty in 1955 where he established a strong research group that rapidly became one of the centers for molecular biology and genetics. He maintained his close links with

James Watson, 1987. *(Courtesy of Cold Spring Harbor Laboratory Archives.)*

Cold Spring Harbor—in 1960 he became a member of the Scientific Advisory Committee and, since 1965, a trustee. This was a critical time for research at Cold Spring Harbor. The Carnegie Institution of Washington withdrew its support for the Department of Genetics, which merged with the Biological Laboratory to form the Cold Spring Harbor Laboratory for Quantitative Biology. There was a reshaping of the Board of Trustees and in 1963, John Cairns came on as director.

It is hard to believe that there were only six full-time laboratories at Cold Spring Harbor in 1963, all, with the exception of McClintock, working on phage and bacterial genetics. The limitation was a severe lack of money, both to hire investigators and to repair an infrastructure that had been neglected for decades. Cairns worked hard and by 1967, he had stabilized the Laboratory's finances, but they continued to be perilous. The effort had taken its toll on him and Cairns gave notice that he would resign as director to return to research. It was at about this time that the Board, knowing of Watson's passion for Cold Spring Harbor, asked him to develop a plan for continued improvement of the Laboratory. Watson did so, and it cannot have surprised him when the Board then offered him the job of director. He negotiated with Harvard that it would continue to pay his salary, even though he would spend one third of his time at Cold Spring Harbor. Watson took up his new post in February 1968.

The change in directorship was accompanied by a marked change in the research interests of the Laboratory, a break with past research that was soon accentuated by, and contributed to, the advent of the new tools of recombinant DNA. The research programs Watson developed as director are each represented by a section in this book. The Tumor Virus program exploited an idea that Watson had had while at Harvard—that the best route to understand what turns a normal cell into a cancer cell was through the study of tumor viruses. By 1980, new techniques had been developed and it became possible to isolate and analyze the oncogenes of mammalian cells. The papers covering this area have here been grouped into the Cancer section. Research in genetics was sustained by Barbara McClintock, still studying transposable elements, and entered the modern era with studies of the mating-type switch and cell cycle in yeast and the transposition of bacteriophage Mu DNA, and moved into genomics when *Arabidopsis* became a favored object of study. The DNA section contains papers on DNA replication and base flipping. Neurobiology came to the Laboratory in 1979 when a new technology, monoclonal antibodies, was used to explore neuronal cell specificity. This fledgling neurobiology program faltered in the early 1980s but was vigorously reinstated in 1990 with studies on the molecular basis of learning and memory in the new Beckman Neuroscience Center. Cell biology was another early program, examining diverse phenomena in cells in tissue culture, some of which related to the Cancer Program. Finally, papers in the Techniques section show how advances in scientific knowledge depend to a degree not usually acknowledged on advances in techniques.

The passage of time meant that it was not possible, with one exception, to ask authors of papers reprinted in *Illuminating Life* 1903–1969 to comment on their work. However, as the authors of papers covered in this volume are still active, and the most interesting insights into the research would come from the researchers themselves, we asked one author from each paper to describe why the experiments came to be done; how they were carried out; what their significance was at the time; and how the research has stood the test of time. We asked, also, for any anecdotes that related to the research and to their time at the Lab.

Each section of this volume is prefaced by a brief introductory essay describing how the papers fit into the Laboratory's research and into the research area in general. We have included photographs of the researchers, contemporary, if possible, with the papers. To keep the printed book to an attractive size, electronic versions of the papers are provided on the enclosed CD.

These papers document how Cold Spring Harbor Laboratory grew from a small institute known for the research of a handful of scientists and for its annual Symposium to a world-class research powerhouse.

JAN A. WITKOWSKI
ALEXANDER GANN
JOSEPH F. SAMBROOK

ACKNOWLEDGMENTS

WE FIRST AND FOREMOST THANK ALL those who contributed essays either recounting their time and work at CSHL, or placing a section of the book in context.

We are very grateful to the staff of the Cold Spring Harbor Laboratory Library and Archives for their help in gathering material for this book, and in particular to Clare Clark for her tireless efforts in tracking down photographs.

Production of the book would not have been possible without the attention and skill of our colleagues at Cold Spring Harbor Laboratory Press, in particular, Maryliz Dickerson who organized and kept track of everything in development, and Mala Mazzullo who took over those tasks in production. Jan Argentine and Denise Weiss oversaw development and production of the book; Carol Brown and Maria Fairchild were our permissions coordinators; Lauren Heller, desktop editor; and Dotty Brown, copy editor.

Reproduction of the papers in this book—and in electronic form on the accompanying CD—was made possible by the generosity of the following publishers, to whom we are most grateful: American Association for the Advancement of Science (publishers of *Science*); American Chemical Society (*Biochemistry*); Cell Press (*Cell*); Elsevier (*Gene*); Rockefeller University Press (*The Journal of Cell Biology*); The National Academy of Sciences (*Proceedings of the National Academy of Sciences*); and Nature Publishing Group (*Nature* and *Nature New Biology*).

GENETICS

Genes Are Interesting to Geneticists

GERALD R. FINK

———

S OON AFTER JIM BECAME DIRECTOR of Cold Spring Harbor
Laboratory in 1968, a guest at one of the many parties held for
local supporters of the Lab asked me, "I know Dr. Watson is a
geneticist; what is he working on now?" This question confronted
me with several problems. First, I wasn't sure Jim was a card-carrying geneticist—I don't think he had ever mated two organisms, the
key move in traditional genetic research. Moreover, once he became
Director, he no longer had a laboratory of his own. Although I don't
remember my exact answer, I muttered a response that gave the illu-
sion of Jim hard at work at something.

As a young man, Jim certainly thought of himself more as a
geneticist than anything else, in that his obsession over the structure
of DNA was a quest to answer the central question of genetics:
"What is a gene?" Comparing himself to Linus Pauling, a chemist,
whose first sentence of a scientific article was, "Collagen is a very
interesting protein," Jim mused,[1] "It inspired me to compose opening
lines of the paper I would write about DNA, if I solved its structure.
A sentence like "Genes are interesting to geneticists" would distinguish my way of thought from Pauling's."

———

GERALD R. FINK, American Cancer Society Professor of Genetics at the Massachusetts Institute of Technology, was a founding member of the Whitehead
Institute, and its Director from 1990 to 2001. He received his Ph.D. in genetics
from Yale University and served for 15 years on the faculty of Cornell University.
His involvement with CSH Laboratory includes teaching on its influential Yeast
Genetics course for 17 years. In recognition, he was awarded one of the first hon-
orary degrees from the Watson School at CSH Laboratory in 1999.

When Jim arrived at Cold Spring Harbor in 1968, classical genetics was represented by the solitary figure of Barbara McClintock, whose purview was not just a single gene, but the entire ensemble of genes that comprise the chromosome complement—what is now called the genome. Her studies on corn genes that moved around the genome raised questions about the stability of genes, their transmission, regulation, and evolution that begged a description in the language of DNA.[2,3] The formality of classical genetics lacked any kinetic aspect that was required for evolution on a genome-wide scale. McClintock shook all that up. She showed that the genome was plastic, constantly being rearranged by transposable elements that could move themselves as well as genes and chromosomes and, by this frenzied activity, dramatically alter gene regulation and genome organization. Three of the four papers included here reflect McClintock's view of the genome, one that was the foundation of much genetic research at CSH for many years.

McClintock desperately sought a DNA-based mechanism for genome rearrangement and constantly plied young molecular biologists for models that could explain her genetic data with corn. Experiments on bacterial transposons translated her genetic revelations into the language of DNA and DNA replication and instantly made McClintock's observations understandable to the new generation of mechanism-oriented molecular biologists. The paper by Ljungquist and Bukhari[4] (see also Harshey, p. 7) is emblematic of the role that bacterial genetics played in uncovering the mechanism that McClintock sought for her corn transposable elements. We now know that some of these elements move by replicating themselves and inserting into novel sites in the genome while others move by cutting themselves out and pasting their DNA into a new site. Moreover, the work of Bukhari and others solidified the connection between these transposable elements and viruses, whose chromosomal insertion in multicellular organisms is responsible for some forms of cancer.

The paper by the CSH Laboratory yeast group (Strathern et al., see also Hicks, p. 13),[5] an energetic troika located in Delbrück (nee Davenport lab), revealed that some DNA movements are programmed, and this molecular traffic is key to development. Yeast cells appeared to have a repository of silent information that could be activated by a directed transposition into another site where it was expressed. This process was not random jumping around the genome àla the McClintock elements, but rather a highly orchestrated event in which a silent cassette of mating-type DNA was moved to another location where it could be expressed (the "playback site"). The net result of this transposition was a programmed change from one mating type to another. And one important consequence of this switch was that a yeast cell could not lead a celibate life: Upon cell division, transposition in a daughter cell of one mating type would produce a cell of the opposite mating type, and these two partners of opposite gender could mate.

The novelty of this mating-type switching caught the imagination of the scientific world because it seemed to contain answers to many perplexing phenomena about gene rearrangement in a wide variety of organisms, for example, the shuffling of genes to create the antibody diversity that protects us from infection and, ironically, the genome rearrangements that permit malaria to switch its cell surface genes, thereby eluding the very antibodies we form to detect that insidious parasite. The CSH yeast group provided direct evidence that the genome contained a reservoir of information that could be activated or silenced during development. Although it turned out that developmental decisions in most organisms do not involve these highly directed genome rearrangements, many genomes do contain regions of silenced genes, and work on how mating-type genes are kept silent in yeast has contributed much to our understanding of the mechanisms involved more generally in silencing. In this sense, the work of the CSH yeast group presaged the current interest in gene silencing, epigenetics, and stem cells.

The realization that transposable elements had the whole genome as their playground led to speculation that these elements might be tamed for experimental uses. So it proved, and they were subsequently employed both as tools for inactivating genes anywhere in the genome and as reporters of gene function. Thus, a mobile genetic element can be engineered to introduce a reporter into any resident gene: The resulting gene fusion reveals the expression pattern, tissue specificity, and potential function of the resident gene. It also inactivates that gene. Traditional mutagenesis procedures provided no special avenue for identifying or isolating the mutated genes. In contrast, insertion mutagenesis (as it is called) enables instant cloning of the inactivated gene because it is tagged with the transposable element DNA.

The paper by Sundaresan et al.[6] (see also Martienssen, p. 19) demonstrated that this transposon strategy could be realized on a genome-wide scale in *Arabidopsis* to detect and clone genes in any plant process. By this time, the Laboratory had embraced *Arabidopsis* instead of corn for molecular genetic studies because of its small size—corn is about the size of a postdoc, whereas *Arabidopsis* on a Petri dish is about the size of a postdoc's fingernail—short generation time, simple genome, and ease of DNA transformation. These features make *Arabidopsis* the ideal laboratory plant. Ironically, the CSH Laboratory plant group used McClintock's corn Ac/Ds transposable elements for these transposition tricks in *Arabidopsis*. The transposon cassette was cleverly designed to obviate many technical and theoretical pitfalls.

Because of the conservation among the genes of plants, *Arabidopsis* genes identified by the strategy elucidated in this paper provided a straightforward way to identify homologous genes in crop plants. *Arabidopsis* genes that affect key traits such as disease resistance, nutrient composition, and yield have thus been used to clone their homologs in corn, rice, and cotton, and these cognates have been used for the subsequent engineering of crop improvement.

These genetically engineered plants have begun to revolutionize agriculture and, the anti-progress lobby notwithstanding, are the hope for our future existence. Genetically modified "golden" rice[7] could prevent the deaths of 350,000 children who die each year and the blindness of another 2 million who lose sight because of vitamin A deficiency. This success in genetically modifying plants has spawned hope that genes identified in *Arabidopsis* will point the way to engineering cellulose in less tractable species such as switchgrass to provide a biofuel alternative to fossil fuels.

It is of course not just between plants that genes for a given process are typically highly conserved through evolution. For example, the assembly of genes that direct intricate developmental events such as cell division have been conserved over eons despite the fact that individual species have often acquired many ornamentations to the process. That is, cell division can look superficially different from one organism to another, but the basic apparatus is the same. Proof of this assertion came from a set of conceptually simple genetic experiments in *Schizosaccharomyces pombe* (often referred to as the British yeast), a fine example of which (Draetta et al.; see also Beach, p. 25)[8] is in the fourth paper in this section.

The genetic approach to understanding cell cycle regulation in yeast was powerful—mutants defective in the process were readily isolated. In each case, a given mutant is unable to divide because it lacks some function required for the process. But what function is it—a kinase, phosphatase, or some unknown chemical reaction? Genetics alone gives no answer. Often, the intricate *biochemistry* of cell division (what enzyme did what to whom) was discovered, not in yeast, but in organisms such as humans, frogs, or clams—organisms that provided lots of cell division enzymes but were minimally amenable to genetics.

So, the trick to figuring out the defect in a mutant yeast strain was to perform molecular miscegenation, i.e., introduce a gene encoding a known biochemical function determined in any organism (human, clam, etc.) into the yeast cell with a cell division defect. If cell division is restored by this marriage, then that biochemical function must play a role in cell division at the point defective in the mutant yeast and likely at exactly the same point in the clam[9] (or whatever) as well. The original version of this experiment (a Nobel-Prize-winning experiment,[10] no less) showed that the human gene encoding a cell cycle protein kinase (Cdc2) when expressed in yeast permitted the otherwise defective yeast strain to complete cell division.

Extending this kind of reasoning, the Draetta et al. paper showed that clam cyclins A and B had an unanticipated biochemical activity—regulation of the Cdc2 protein kinase activity cyclically through the cell cycle. These experiments not only elaborated the critical sequence of events in the cell division pathway, but also reasserted strongly the important evolutionary statement: Despite the fact that cell division looks visibly different from one species to another, the basic clockwork has been conserved in each genome, whether they are yeast or clams or humans.

The question that obsessed Jim Watson as a youth—"What is a gene?"—has resurfaced as a fundamental question in biology today. There are now prions, microRNAs, and "junk DNA," which do not fit the classical definition of a gene but are nonetheless transmitted from one generation to the next. Their chemical composition, transmission, and function form the obsessions of the next generation. Genes are still interesting to geneticists.

Notes and References

1. Watson J.D. 1968. *The Double Helix.* Atheneum, New York.

2. Fink G.R. 1992. Barbara McClintock (1902–1992). *Nature* **359:** 272.

3. Fedoroff N. and Botstein D., eds. 1992. *The dynamic genome: Barbara McClintock's ideas in the century of genetics.* Cold Spring Harbor Laboratory Press, Cold Spring Harbor, New York.

4. Ljungquist E. and Bukhari A. 1977. State of prophage Mu DNA upon induction. *Proc. Natl. Acad. Sci.* **74:** 3143–3147.

5. Strathern J.N, Klar A.J., Hicks J.B., Abraham J.A., Ivy J.M., Nasmyth K.A., and McGill C. 1982. Homothallic switching of yeast mating type cassettes is initiated by a double-stranded cut in the *MAT* locus. *Cell* **31:** 183–192.

6. Sundaresan V., Springer P., Volpe T., Haward S., Jones J.D.G., Dean C., Ma H., and Martienssen R. 1995. Patterns of gene-action in plant development revealed by enhancer trap and gene trap transposable elements. *Genes Dev.* **9:** 1797–1810.

7. Ye X., Al-Babili S., Klöti A., Zhang J., Lucca P., Beyer P., and Potrykus I. 2000. Engineering the provitamin A (beta-carotene) biosynthetic pathway into (carotenoid-free) rice endosperm. *Science* **287:** 303–305.

8. Draetta G., Luca F., Westendorf J., Brizuela L., Ruderman J., and Beach D. 1989. Cdc2 protein kinase is complexed with both cyclin A and B: Evidence for proteolytic inactivation of MPF. *Cell* **56:** 829–838.

9. Swenson K.I., Farrell K.M., and Ruderman J.V. 1986. The calm embryo protein cyclin A induces entry into M phase and the resumption of meiosis in *Xenopus* oocytes. *Cell* **47:** 861–870.

10. Lee M.G. and Nurse P. 1987. Complementation used to clone a human homologue of the fission yeast cell cycle control gene *cdc2*. *Nature* **327:** 31–35.

Ljungquist E. and Bukhari A.I. 1977. **State of prophage Mu DNA upon induction.** *Proc. Natl. Acad. Sci.* **74:** 3143–3147. (Reprinted, with permission.)

THIS PAPER EXPOUNDS TWO NOTIONS, startling for their time: A segment of DNA can transpose from one chromosomal location to another without it leaving the original location, and replicas flit to new sites in the genome without having a discernible independent existence.

The idea of genetic elements within the host genome goes back the 1950s—the heyday of cytogenetics. Working at Cold Spring Harbor, Barbara McClintock proposed[1] that gene activity in maize may be regulated by transposable elements that can insert into, and be excised from, different chromosomal loci. The molecular nature of these elements was unknown until the discovery of insertion elements in *Escherichia coli*, the first example of which was Mu, the "mu"tagenic bacteriophage. In 1963, Austin Taylor reported[2] that Mu lysogens acquired a wide range of mutations, each of which was genetically inseparable from an integrated copy of Mu DNA, leading him to infer that Mu can integrate at a large number of sites in the *E. coli* genome. It was the ability of Mu to cause mutations promiscuously that aroused Ahmad Bukhari's interest when, in the mid 1960s, he became Taylor's first graduate student at the University of Colorado at Denver. Although Mu was not the topic of Bukhari's dissertation, he became entranced by the bacteriophage and took it with him to Cold Spring Harbor when he joined David Zipser's[3] laboratory as a postdoctoral fellow in 1970. Around this time, reports of other insertion elements in *E. coli* were beginning to be published.[4]

By 1972, Bukhari had established his own laboratory at Cold Spring Harbor, which soon became the focal point for work on bacteriophage Mu. The first of Ahmad's many pioneering contributions was the demonstration that Mu could insert randomly at many sites within a single bacterial gene.[5] In a monumental

---

RASIKA HARSHEY holds a Ph.D. from the Indian Institute of Science, Bangalore. She joined Cold Spring Harbor in the late 1970s as a postdoctoral fellow in Ahmad Bukhari's laboratory and was soon promoted to the scientific staff. After working at the Laboratory for two more years, she took a position at the Scripps Research Institute, La Jolla. She is currently a Professor in the Section of Molecular Genetics and Microbiology at the Austin campus of the University of Texas. rasika@uts.cc.utexas.edu

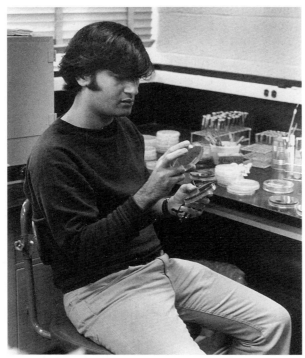

Ahmad Bukhari, 1973. *(Photograph by Ross Meurer, courtesy of Cold Spring Harbor Laboratory Archives.)*

piece of work, Bukhari showed that these insertions were extremely stable.[6] On the rare occasions that they reverted, restoration of gene activity was accompanied by excision of the integrated prophage, a formal proof that Mu integration and excision could occur without damaging the target DNA sequences. Demonstrations for other transposable elements soon followed and the consimilar behavior of Mu, IS sequences, and other transposons was the stimulus for a meeting coorganized by Bukhari at Cold Spring Harbor in the summer of 1976,[7] which was pivotal in accelerating research into these elements.

Bukhari was convinced that a solution to the mechanism of transposition would come from Mu, primarily because of the relatively rapid movement of the bacteriophage genome around the *E. coli* chromosome: Mu transposes at a rate a hundred thousand times higher than other movable genetic elements. However, well into the 1970s, Mu was sometimes dismissed as a promiscuous maverick by the more traditional members of the bacteriophage community, who believed that the mechanism of integration/excision had already been satisfactorily solved by research on bacteriophage λ—the old warhorse of bacteriophage genetics. During integration of λ, the "sticky ends" of the bacteriophage DNA join together to generate circular molecules that integrate into a single specific target site in the *E. coli* chromosome by recipro- cal recombination. Conversely, reversal of the recombination event—again by recombination—liberates a free copy of λ DNA from its site in the bacterial genome. In contrast, Mu DNA does not have cohesive ends, does not make covalent circles, does not integrate at a single site, and does not generate replicating intermediates like λ. The paper reproduced here not only settled the "is Mu like λ" question defin- itively, but also marked a turning point in thinking about transposition. Using restriction analysis and Southern DNA hybridization, Ljungquist and Bukhari first

induced Mu and λ prophages to enter into lytic growth and then followed the fate of the prophage-host DNA junctions for the next 30 minutes. The original λ-host junctions disappeared soon after induction, consistent with physical excision of λ prophage. However, the original Mu-host junc- tions remained intact late into the lytic cycle, concomitant with the appearance of many new Mu-host junctions. Because Mu was replicating during this time, it was apparent that Mu could move to new sites without leaving its original location. Thus, unlike λ, parental Mu was not excised.

Rasika Harshey, 1979. *(Photograph by Ross Meurer, courtesy of Cold Spring Harbor Laboratory Archives.)*

The nonexcisive, concerted rep- lication-integration mechanism for Mu transposition suggested by the Ljungquist- Bukhari experiment served as the archetype for many subsequent transposition models. However, as more transposable elements were analyzed, it became clear that the type of replicative transposition used by Mu is shared by only a handful of genetic elements. Most, like those described by McClintock in maize, excise and reinsert without replication. Repair of the chromosomal breaks left in their wake often involves invasion of the copy of the element integrated into the sister chro- matid, a process that can give the impression of replication. Although Mu can no longer be regarded as the prototype of movable genetic elements, the bacterio- phage was the catalyst that liberated thinking about transposition, its mechanisms and significance. Many of the current ideas on movement of DNA segments from one chromosomal location to another are the fruits of this liberation.

Ahmad Bukhari loved Mu with a passion and promoted it indefatigably. His enthusiasm was infectious. He always saw his work in a larger context and was instrumental in ushering in a new era of research on movable genetic elements during his brief tenure at Cold Spring Harbor. He died of a heart attack while jogging on November 19, 1983. He was 40 years old.

Notes and References

1. McClintock B. 1950. The origin and behavior of mutable loci in maize. *Proc. Natl. Acad. Sci.* **36:** 344–355.

2. Taylor A.L. 1963. Bacteriophage-induced mutation in *Escherichia coli. Proc. Natl. Acad. Sci.* **50:** 1043–1051.

3. David Zipser moved to Cold Spring Harbor from Columbia University in 1970 and, as Jim Watson wrote in his annual report for that year, "With [Zipser's] arrival, the laboratory regains solid strength in bacterial genetics, an area in which it held a commanding presence for over two decades." Zipser's laboratory maintained this tradition for the next 12 years, but, by the beginning of the 1980s, the attraction of *E. coli* as a model genetic system had begun to fade. The center of gravity of the Laboratory had swung irrevocably toward mammalian cells and their viruses. Zipser's interests had already shifted toward computational analysis of biological systems, and in 1982, he left the Laboratory to join the Department of Cognitive Science at the University of California, La Jolla.

4. Described in the *Introduction* and cataloged in *Appendix A* of *DNA insertion elements, plasmids, and episomes* (1977. ed. Bukhari et al., Cold Spring Harbor Laboratory, Cold Spring Harbor, New York).

5. Bukhari A.I. and Zipser D. 1972. Random insertion of Mu-1 DNA within a single gene. *Nat. New Biol.* **236:** 240–243.

6. Bukhari A.I. 1975. Reversal of mutator phage Mu integration. *J. Mol. Biol.* **96:** 87–99.

7. The meeting, *DNA Insertions,* was organized by Ahmad Bukhari, James Shapiro (University of Chicago), and Sankar Adhya (National Institutes of Health).

8. Craig N.L., Craigie R., Gellert M., and Lambowitz A. 2002. *Mobile DNA II.* American Society for Microbiology, Washington, D.C.

Proc. Natl. Acad. Sci. USA
Vol. 74, No. 8, pp. 3143–3147, August 1977
Biochemistry

State of prophage Mu DNA upon induction

(bacteriophage Mu/bacteriophage λ/DNA insertion/DNA excision/transposable elements)

E. LJUNGQUIST AND A. I. BUKHARI

Cold Spring Harbor Laboratory, Cold Spring Harbor, New York 11724

Communicated by J. D. Watson, April 18, 1977

ABSTRACT   We have compared the process of prophage λ induction with that of prophage Mu. According to the Campbell model, rescue of λ DNA from the host DNA involves reversal of λ integration such that the prophage DNA is excised from the host chromosome. We have monitored this event by locating the prophage DNA with a technique in which DNA of the lysogenic cells is cleaved with a restriction endonuclease and fractionated in agarose gels. The DNA fragments are denatured in gels, transferred to a nitrocellulose paper, and hybridized with ^{32}P-labeled mature phage DNA. The fragments containing prophage DNA become visible after autoradiography. Upon prophage λ induction, the phage–host junction fragments disappear and the fragment containing the λ att site appears. No such excision is seen in prophage Mu. The Mu–host junction fragments remain intact well into the lytic cycle, when Mu DNA has undergone many rounds of replication and apparently many copies of Mu DNA have been integrated into the host DNA. Therefore, we postulate that Mu DNA replicates in situ and that the replication generates a form of Mu DNA active in the integrative recombination between Mu DNA and host DNA. This type of mechanism may be common to many transposable elements.

Assimilation of one DNA molecule into another is a fundamental biological phenomenon spanning the whole range of prokaryotic and eukaryotic organisms. The insertion problem has been studied in its most clear-cut form in viral systems. The classic mode of integration of viral genomes involves recombination between the host DNA and a circular form of the inserting viral DNA. Specific sequences in the inserting circular genomes and in the host DNA are recognized for recombination, so that the process culminates in complete linear insertion of the circles (1). Rescue of the inserted DNA from the host is visualized as physical excision of the DNA in a manner that is a reversal of the insertion process. The question of whether or not the temperate bacteriophage Mu conforms to this classic mode of integration and excision has been the focus of the current work on Mu (2). This question has arisen because Mu is strikingly different from other temperate viruses of bacteria in several features.

Unlike other temperate phages, Mu inserts its DNA at randomly distributed sites on the genome of its host bacterium Escherichia coli (3, 4). As extracted from mature phage particles, Mu DNA, is a linear duplex of 37–38 kilobases and has at its ends host DNA that differs in size and sequence from molecule to molecule (5, 6). Thus, Mu does not have terminally cohesive or repetitious sequences and lacks any obvious means of fusing its ends to form circular DNA molecules. The terminal host sequences are randomized during Mu growth, because Mu lysates grown from a single plaque still contain particles with

different host sequences. Yet, a form of Mu DNA free of host DNA has remained undetected.

In its continuous association with host DNA, Mu resembles another class of insertion elements, referred to as the transposable elements. The transposable elements are specific stretches of DNA that can be translocated from one position to another in host DNA (7). Mu undergoes multiple rounds of transposition during its growth, far exceeding the transposition frequency of the bona fide transposable elements. When a Mu lysogen, carrying a single Mu prophage at a given site on the host chromosome, is induced, many copies of Mu DNA are rapidly integrated at different sites as the replication of Mu DNA proceeds.

We have sought to determine whether induction of a Mu prophage, with subsequent replication and transposition of Mu DNA, involves excision of the prophage DNA from the original site. To do this, we have examined the fate of prophage Mu DNA, and also of prophage λ DNA, in situ in the host chromosome after induction. This paper presents evidence that, unlike λ, prophage Mu DNA persists at its original site after induction.

MATERIALS AND METHODS

Bacterial Strains. The E. coli strains were all derivatives of E. coli K-12. The basic bacteriophages in the lysogenic cells were either Mu cts62, a temperature-inducible mutant of bacteriophage Mu carrying a mutation in the immunity gene c (8), or λ cI857S7, a temperature-inducible derivative of bacteriophage λ. The Mu cts62 lysogens were: BU563 (Mu cts62 located in one of the pro genes), BU568 (Mu cts62 located at the thr locus), BU575 (Mu cts62 located at the trp locus), BU8220 (Mu cts62 located in the lacI gene on an F' pro+ lac episome). BU1216 carried the Mu A gene mutant Mu cts62 Ats5045. The λcI 857S7 lysogen was BU851.

Genetic Procedures. The media, growth, and induction conditions have been described in detail by Bukhari and Ljungquist (9).

Biochemical Procedures.

(i) Extraction of DNA. The bacterial cells were washed, resuspended in 0.01 M Tris·HCl/1 mM EDTA, pH 7.9, and lysed by the addition of 0.5% sodium dodecyl sulfate. The lysate was digested with Pronase (self-digested for 2 hr at 37° in 0.01 M Tris·HCl, pH 7.4) at a concentration of 1 mg/ml for 8 hr. The solution was then extracted twice with Tris buffer/EDTA/saturated phenol. The aqueous phase was dialyzed against 0.01 M Tris·HCl/1 mM EDTA, pH 7.9, after which it was treated with RNase at a concentration of 100 μg/ml for 2 hr at 37° and then with Pronase at 100 μg/ml for 3 hr at 37°. The DNA was extracted with phenol again and dialyzed as above.

Abbreviations: EcoRI, Bgl II, and Bal I refer to restriction endonucleases from E. coli RY 13, Bacillus globiggi, and Brevibacterium albidum, respectively.

In Memoriam

Ahmad I. Bukhari, 1943–1983

On November 19, 1983 we lost our dear friend, Ahmad Iqbal Bukhari, one of the Editors of this journal. While jogging on Saturday morning near his home in Huntington, Long Island, NY, he suffered a fatal heart attack. He was only 40 years old, a very productive and innovative scientist, pioneering and deeply involved in the molecular genetics of bacteriophage Mu and transposing elements, inveterate organizer of phage meetings, as well as a very warm human being.

Ahmad, a naturalized U.S. citizen, was born on January 5, 1943 in Punjab, India, which later became a part of Pakistan. He received his B.Sc. and M.Sc. degrees from the University of Karachi, and came to the United States as a Fulbright-Hays scholar at Brown University, where he received the M.S. degree in 1966. He earned his Ph.D. degree under Prof. A.L. Taylor at the University of Colorado Medical Center, Denver, CO, performing research on the biochemical genetics of diaminopimelic acid and lysine synthesis in *Escherichia coli*. During this time he used bacteriophage Mu to generate auxotrophic mutants in the lysine biosynthetic pathway and became keenly interested in trying to decipher the life cycle of this unusual phage.

After leaving Colorado for a postdoctoral position in Prof. D. Zipser's laboratory at Cold Spring Harbor Laboratory, Ahmad vigorously pursued his long-range goal of solving the riddles surrounding phage Mu, and while doing so inspired many others to join the family of Mu researchers. In the course of his studies as a Jane Coffin Childs Fellow, and subsequently as a Staff Investigator and Laboratory Chief at Cold Spring Harbor, Ahmad and his many working colleagues made numerous discoveries of great fundamental importance to the current understanding of phage Mu as a transposable genetic element. For those who work on phage Mu and other transposable elements, it can be said that Ahmad's keen scientific insight and tireless devotion to doing the right experiment will be sorely missed in the years ahead.

Aside from his own research, Dr. Bukhari was deeply committed to helping the international scientific community. Toward this end, he served since 1981 as an expert advisor to the United Nations International Development Organization (UNIDO) to help establish an International Center for Genetic Engineering and Biotechnology. The aim of the center, which is now in the final stages of planning, is to increase the flow of biotechnology to developing countries.

Dr. Bukhari was also a scientific advisor to the United Nations Educational, Scientific and Cultural Organization (UNESCO), and played a key role in establishing a national biotechnology center for Pakistan, located in Lahore.

The author of numerous scientific articles, Dr. Bukhari was a member of the American Society for Microbiology, the Genetics Society of America, and the American Association for the Advancement of Science.

Through his fostering of workshops, symposia and other meetings, Dr. Bukhari inspired a great many investigators. As a consequence, numerous scientists are now engaged in expanding knowledge in the areas of his activities. The Cold Spring Harbor Phage Meetings were very close to his heart. His last activity was organization of the Fourth International Bacteriophage Mu Workshop in his native Pakistan, held December 5 – 8, 1983 in Lahore, just after his death, and dedicated to his memory.

Dr. Bukhari is survived by his wife, Christine K. Morgan, a researcher for Time Magazine and by two sons, Yousaf Ali, 3, and Jaffet Ahmad, 6 months; his father, Noor-Ul-Islam Bukhari and a brother, Ahmad Abid, in Pakistan; a brother, Asad, in Lloyd Harbor, NY; and a sister, Shahida Jalal, in Rocky Point, NY. A memorial fund for his children has been set up with contributions to be sent to Robin Wilson, Cold Spring Harbor Laboratory, P.O. Box 100, Cold Spring Harbor, NY 11724 (U.S.A.).

Austin L. Taylor
Department of Microbiology and Immunology
University of Colorado Health Sciences Center
Denver, CO 80262 (U.S.A.)
Tel. (303) 394-7903

Waclaw Szybalski
Editor-in-Chief

Strathern J.N., Klar A.J.S., Hicks J.B., Abraham J.A., Ivy J.M., Nasmyth K.A., and McGill C. 1982. **Homothallic switching of yeast mating type cassettes is initiated by a double-stranded cut in the *MAT* locus.** *Cell* **31:** 183–192. (Reprinted, with permission, from Elsevier ©1982.)

I N 1977, WHEN I WAS A POSTDOC in Gerry Fink's laboratory at Cornell, Ray Gesteland,[1] then the Assistant Director at Cold Spring Harbor, asked me what I wanted in the way of a job. I said that I didn't really want a job yet, but that I really wanted to hook up with my graduate school partner Jeff Strathern and validate the "Cassette Model" for yeast mating-type genes that we had formulated in Ira Herskowitz's lab at Oregon. I had just the place in mind, the Davenport teaching lab at Cold Spring Harbor. It was set up for yeast genetics but was only used in the summer months for courses. Ray thought that might be possible, and even exciting, but he was worried when I told him that Jeff and his wife, Anne, raised golden retrievers. "Jim's death on dogs," he said.

Dogs notwithstanding, Ray seemed to be able to arrange things and Jeff agreed to put off his planned postdoc in worm genetics (as it turns out, Jim Watson was more interested in establishing a yeast genetics lab at Cold Spring Harbor than in worrying about pet irritations), and soon we were writing grant applications to fund our adventure. We spent many hours designing baroque strategies for cloning the mating-type genes based on the structures we had deduced from genetic studies, but when Albert Hinnen and I achieved yeast transformation in Gerry's lab in 1978,[2] the way became clear. We would clone mating type by functional complementation of mutants.

By then, however, another player had come into our lives. The irrespressible Amar Klar[3] turned up at the 1977 Cold Spring Harbor Yeast Meeting with aston-

JIM HICKS, PH.D., has served as a Director of AVI since 1997. He is Senior Invetigator and Director of the Women's Cancer Genomics Initiative at Cold Spring Harbor Laboratory. He has served as the Director and Chief Technology Officer of Virogenomics, Inc., a biotechnology company, since 2001. Dr. Hicks currently serves on the board of Barrett Business Services. He received his B.A. degree in Biology from Willamette University and his Ph.D. in Molecular Biology from the University of Oregon, followed by postdoctoral research at Cornell University. hicks@cshl.edu

Jim Hicks, Amar Klar, Jeff Strathern, 1980 Yeast Lab Group. *(Photograph by Joan James, courtesy of Cold Spring Harbor Laboratory Archives.)*

ishing genetic data on mating type that complemented our own. We were first a little worried about the unexpected competition, but Ray said why not just have him join the group here? Amar was agreeable and we converged on Davenport Lab in the fall of 1978 for our first nine-month season of frantic genetics and molecular biology and brainstorming and fun.

The genetics of mating type eventually left its stamp on many areas of biology. As diagrammed in the accompanying paper, the mating-type genes of *Saccharomyces* yeast are arrayed in three distinct loci on a single chromosome. All three loci contain functional genes, but two of them, located near the telomeres, are kept silenced by *trans*-acting factors produced by the SIR genes acting through modification of chromatin structure. *SIR2* (also known as "*a"MAR1*) eventually gave its name to the mammalian genes known as "sirtuins." But that gets ahead of the story.

Long before the physical arrangement of the yeast mating-type genes was known, their organization was predicted from genetic mapping data by Takano and Oshima in 1971[4] and the big biological question was how a developmental process called mating-type switching could be possible. Most laboratory strains (called ho) grow as haploids and exhibit a mating type either **a** or α. When allowed to mate, cells of opposite type exchange sex pheromones and prepare to mate, forming a diploid **a**/α cell type that can also grow stably in the lab. They are generally diploid and grow stably until they starve. At that point, they are signaled to undergo meiosis, giving rise to four haploid spores, presumably to better survive winter in the wild. Each of the spores exhibits one of the two mating types, either **a** or α and the two types segregate as alleles, two **a** and two α, exactly as Mendel predicted. Wild yeast and certain laboratory strains, however, carry a gene called *HO*. If a spore from such a strain is separated either by wind in the wild or by laboratory manipulation, it exhibits surprising behavior. Within two generations, each single haploid spore of either mating type switches, yielding a few cells of the opposite mating type. This occurs in a formal pattern that Jeff and I catalogued when we were graduate students. If left alone, these cells of opposite mating type can fuse and once again create the stable *HO* diploid. It struck both of us that there was clearly an organized developmental process under way that facilitated the alteration of generations and we wanted to follow it to the end. Such switches are the essence of development in all organisms.

Switching was intriguing by itself, but another result led us to devote many years to this system. Both Jeff and I at Oregon[5] and Amar at Berkeley[6] independently obtained the profound result that genetic material was actually transferred from one place to another during switching. At Oregon, we showed that an active, but defective, mating-type gene could be "healed" by switching, presumably with DNA from another site. Unknown to us, Amar had performed the converse experiment, "wounding" the MAT locus by inserting a defective mating-type allele from one of the ancillary sites. Always at ease with simplifying concepts, Ira coined the term "Cassette Model" for this phenomenon.

On the basis of the "healing" and "wounding" results, coupled with our colleagues Takano and Oshima's genetic mapping data results, we predicted that the ancillary genes were exact copies of the *MAT* genes and that switching occurred by copying actual DNA information from one site to another. Our job at Cold Spring Harbor was to understand the mechanics of that process. We were able to confirm not only the arrangement, but also the structure of the genes by cloning all three loci in 1979 and comparing them by electron microscopy. Our exciting exploration of the details of that simple, yet profound, switch is exemplified in the accompanying 1982 paper.

Notes and References

1. Ray Gesteland, who had been a Ph.D. student of Jim Watson's at Harvard, was Jim's first appointment after he took over as Director of Cold Spring Harbor in 1967. Ray became Jim's right-hand person and for ten years was responsible for most of the scientific administration of the Laboratory. Ray left Cold Spring Harbor in 1987 to join the University of Utah as the Howard Hughes Professor of Biology.

2. Hinnen A., Hicks J.B., and Fink G.R. 1978. Transformation of yeast. *Proc. Natl. Acad. Sci.* **75:** 1929–1933.

3. Amar Klar, Jeff Strathern, and Jim Hicks fulfilled Jim Watson's ambition for a thriving year-round yeast research laboratory at Cold Spring Harbor. In the 1970 edition of *The Molecular Biology of the Gene*, Jim wrote

> There are now many reasons to intensify work on organisms like yeast. The very concentrated effort which has gone into the study of all aspects of *E. coli* is one of the major reasons why molecular biology has advanced so rapidly over the past two decades. Clearly, similar attention will soon be placed on one or more types of eucaryotic cells. For many reasons it is natural that much emphasis must go toward the study of several types of human cells. But at the same time, it may be wise to concentrate equally on the molecular biology of one or two of the simplest eucaryotes. Several reasons dictate this approach. One is that these microorganisms most certainly contain much less DNA than human cells. Only a five- to tenfold increase in genetic complexity is noticed in escalating to yeast or *Aspergillus* from *E. coli*. A second reason is economic: work with higher cells is at least an order of magnitude more expensive than with microorganisms. If a choice exists between solving the problem with human tissue culture cells or with yeast, common sense tells us to stick with the simpler system. A third, and perhaps the most important reason, is the ease with which detailed genetic analysis can be applied to many microorganisms. Despite the great advantages now brought about by the cell-fusion technique, detailed genetic analysis of human cells will be extraordinarily difficult to bring about. Thus, even if our primary interest is the human cell, this may be the time for many more biologists to work with organisms like yeasts.

The yeast group was run by a triad: Jeff Strathern, who had obtained his Ph.D. from the University of Oregon; Amar Klar, who had worked as a postdoctoral fellow with Seymour Fogel at the University of California; and Jim Hicks, who had been a postdoctoral fellow with Gerry Fink at Cornell. The yeast group worked at the Lab for seven years, and their work during that time is taught in every textbook of genetics and cell biology. However, not long after solving the problem of mating-type switching, the group disbanded. Amar Klar and Jeff Stratern moved to the Frederick campus of the National Cancer Institute. Jim Hicks went to the west coast, first to the Scripps Institute in La Jolla and afterward to ICOS Corporation and Hedral Therapeutics, two biotechnology companies. Most recently, Jim has returned to Cold Spring Harbor Laboratory to join Mike Wigler's group, working on genetic abnormalities in breast cancers.

4. Oshima Y. and Takano I. 1971. Mating types in *Saccharomyces*: Their convertibility and homothallism. *Genetics* **67:** 327–335.

5. Hicks J. and Herskowitz I. 1976. Interconversion of yeast mating types II. Restoration of mating ability to sterile mutants in homothallic and heterothallic strains. *Genetics* **85:** 373–393

6. Klar A., Fogel S., and Radin D. 1979. Switching of a mating type a mutant allele in budding yeast *Saccharomyces cerevisiae. Genetics* **92:** 759–776.

Cell, Vol. 31, 183–192, November 1982, Copyright © 1982 by MIT

Homothallic Switching of Yeast Mating Type Cassettes Is Initiated by a Double-Stranded Cut in the *MAT* Locus

Jeffrey N. Strathern,* Amar J. S. Klar,*
James B. Hicks,* Judith A. Abraham,*
John M. Ivy,* Kim A. Nasmyth[†] and
Carolyn McGill*
* Cold Spring Harbor Laboratory
Cold Spring Harbor, New York 11724
† Medical Research Council
Hills Road
Cambridge CB2 2QH, England

Summary

A double-stranded DNA cut has been observed in the mating type (*MAT*) locus of the yeast Saccharomyces cerevisiae in cultures undergoing homothallic cassette switching. Cutting is observed in exponentially growing cells of genotype *HO HMLα MAT*α *HMR*α or *HO HML*a *MAT*a *HMR*a, which switch continuously, but not in a/α *HO/HO* diploid strains, in which homothallic switching is known to be shut off. Stationary phase cultures do not exhibit the cut. Although this site-specific cut occurs in a sequence (Z1) common to the silent *HML* and *HMR* cassettes and to *MAT*, only the Z1 sequence at the *MAT* locus is cut. The cut at *MAT* occurs in the absence of the *HML* and *HMR* donor cassettes, suggesting that cutting initiates the switching process. An assay for switching on hybrid plasmids containing *mat*a⁻ cassettes has been devised, and deletion mapping has shown that the cut site is required for efficient switching. Thus a double-stranded cut at the *MAT* locus appears to initiate cassette transposition–substitution and defines *MAT* as the recipient in this process.

Introduction

Differentiation of cell type in Saccharomyces cerevisiae is mediated by a specific DNA transposition–substitution event. The a and α mating types of Saccharomyces yeasts are controlled by the *MAT*a and *MAT*α alleles of the mating type locus (Hawthorne 1963a; MacKay and Manney 1974a, 1974b; Strathern et al., 1981). These alleles are genetically codominant and differ by a DNA substitution (Figure 1). *MAT*a and *MAT*α are inherited as stable alleles in strains designated heterothallic (strains carrying the *ho* mutation). Mating of an a haploid with an α haploid produces an a/α diploid that does not mate but is capable of meiosis and sporulation, regenerating a and α haploid cells. In homothallic yeast (strains with the *HO* allele) interconversions between the *MAT*a and *MAT*α alleles occur by transposition and substitution of copies of unexpressed a or α sequences from storage sites elsewhere in the genome into the mating type locus—the cassette mechanism of mating type switching (Hicks et al., 1977a). The storage sites

for the unexpressed donor sequences *HML* and *HMR* have been identified genetically (Takano and Oshima, 1967; Rine et al., 1979; Klar et al., 1979) and physically (Hicks et al., 1979; Nasmyth and Tatchell, 1980; Strathern et al., 1980). As shown in Figure 1, *MAT*, *HML* and *HMR* are located on chromosome III (Harashima and Oshima, 1976; Klar et al., 1980a). *HMR* is the homothallism locus on the right arm of chromosome III as it is conventionally drawn, whereas *HML* is on the left arm. *HML* and *HMR*, like *MAT*, can contain either the a-specific (*HML*a, *HMR*a) or the α-specific (*HML*α, *HMR*α) sequences. In addition to the a-specific or α-specific sequences, *HML* and *HMR* share other regions of homology with *MAT*, as described in the legend to Figure 1.

HML and *HMR* contain complete copies of the *MAT*a or *MAT*α genes, but they are normally not expressed. The sequences stored at *HML* and *HMR* are kept silent by the action of trans-acting negative regulators that are the products of four unlinked genes *SIR1*, *MAR1* (*SIR2*), *SIR3* and *SIR4* (Rine et al., 1979; Klar et al., 1979; Rine, 1979; Haber and George, 1979). In Sir⁻ or Mar⁻ cells the cassettes stored at *HML* and *HMR* are expressed and provide the same regulatory functions normally provided by *MAT*. The mechanism by which the *MAR/SIR* products repress transcription of *HML* and *HMR* is the subject of another manuscript from this laboratory (Nasmyth, 1982). The *MAR/SIR* functions also control the donor versus recipient roles of *HML* and *HMR*. That is, in Mar⁺/Sir⁺ cells, *HML* and *HMR* act as donors in the switching process, whereas in Mar⁻ cells they can also act as recipients (Klar et al., 1981a). We address the mechanism by which copies of the cassettes at *HML* or *HMR* are substituted into *MAT*.

Results

In Strains Undergoing Switching, A Double-Stranded Cut Exists in the *MAT* DNA

To visualize intermediates in the switching process, it was necessary to identify conditions under which a population of continuously switching cells could be grown. Most homothallic strains have the genotype *HO HMLα HMR*a. In these strains, haploid a cells can use *HML*α to switch to *MAT*α, and α cells can use *HMR*a to switch to *MAT*a. Subsequently, sibling cells of opposite mating type mate to form a/α diploids. Switching ceases in these a/α cells. This probably reflects the observation that the *HO* gene is not transcribed (R. Jensen and I. Herskowitz, personal communication). As a result, it is difficult to maintain a population of switching *HO HMLα HMR*a cells, since such cultures are predominantly composed of nonswitching a/α cells.

In contrast, *HO HML*a *HMR*a *MAT*a cells (designated Hqa; Takano and Oshima, 1970) appear to be stable clones of a cells. In reality, homothallic switch-

ROBERT MARTIENSSEN

Sundaresan V., Springer P., Volpe T., Haward S., Jones J.D.G., Dean C., Ma H., and Martienssen R. 1995. **Patterns of gene action in plant development revealed by enhancer trap and gene trap transposable elements.** *Genes Dev.* **9:** 1797–1810. (Reprinted, with permission, ©1995 by Cold Spring Harbor Laboratory Press.)

I am sure now that we can get many more newly arising mutable *a1* loci. The method of detection is simple. In fact, I think that we can go into business. If anyone wants a locus to be mutable, just put in the order and one will be sent the following year! This is not as facetious as it may sound, for that is the way it is turning out and now I know how to spot them rapidly.

Barbara McClintock, 2nd September 1950, in a letter to
Marcus Rhoades (courtesy N. Comfort)

I INTERVIEWED FOR THE VERY JUNIOR FACULTY POSITION of "Staff Investivator" at Cold Spring Harbor Laboratory in the summer of 1988. I had been persuaded to apply to the Lab by Venkatesan Sundaresan (Sundar) my friend and former colleague from Mike Freeling's lab at Berkeley. In those days, plant genetics candidates attracted a small audience, comprising in my case only six people, but as these included Jim Watson, Barbara McClintock, and Rich Roberts, it was a daunting group. Reaction to my presentation on maize transposons was mixed, but Barbara was enthusiastic, gently correcting my mistakes and encouraging me to come to CSH Laboratory.

Shortly after I arrived in January 1989, I was struggling to sequence a transposon-tagged gene in maize. Transposon tagging was a powerful approach to identifying genes but was only widely available at the time in *Drosophila*. Unfortunately, my

---

ROB MARTIENSSEN is a plant geneticist, working on maize and the model plant *Arabidopsis thaliana*, and on RNA interferance in yeast. He received his B.A. in Natural Sciences (Genetics) from Cambridge University, England, in 1982 and his Ph.D. from the Plant Breeding Institute and Cambridge University in 1986. After postdoctoral training at the University of California at Berkeley, he joined the faculty at Cold Spring Harbor Laboratory in 1989 and is currently the Head of Plant Genetics. martiens@cshl.edu

Rob Martienssen and Barbara McClintock, circa 1990. *(Photograph by Tim Mulligan, courtesy of Cold Spring Harbor Laboratory Archives.)*

first attempts at sequencing were unsuccessful, and just as I realized this, Jim walked into the lab, at 10 o'clock on a Sunday morning. Unprepared, I was reassured at once. "Don't worry," Jim said, "we are going to sequence *Arabidopsis*." "Wonderful," I thought, without perhaps appreciating what was meant by "we." *Arabidopsis thaliana* was soon to become the *Drosophila* of the plant world. Many years and 115 Mb later, I understood that first we had to sequence the genome of *Arabidopsis* and then come up with a scheme for tagging. Needless to say, all was accomplished, although not necessarily in this order. The initiative to sequence *Arabidopsis* began during the Cold Spring Harbor Plant Course two years later. Jim drove Sundar, myself, Joanne Chory, Joe Ecker, Mike Bevan, and Sakis Theologis to lunch at Canterbury Ales in Oyster Bay. Jim had to drop off books at the library before lunch, leaving us to swelter in the summer heat, shoulder to shoulder, in his station wagon, waiting to hear what Jim had in store. The Arabidopsis Genome Project was founded over fish and chips.

The inspiration for enhancer detection in *Arabidopsis* came from studies in *Drosophila*. My old friend Kevin Moses had extolled its virtues, as had David Bowtell, in Gerry Rubin's lab at Berkeley in the 1980s, and it was easy enough to apply to *Arabidopsis*. Using ideas from Jonathan Jones, Caroline Dean, and Bill Skarnes, we made constructs using the transposons that Barbara had first discovered—Ac/Ds in maize. Fitted with *beta glucuronidase* or *"GUS"* (the *"plant lacZ"*), we expected that transpositions of these constructs into plant genes would show gene expression patterns as well as causing a mutant phenotype.

Sundar and I divided up the work: Sundar made the gene-trap splice sites and tested them in transient expression experiments using the famous "gene gun," which led to a number of jokes in the lab—shotgun libraries took on a whole new meaning. I made the positive-negative selection scheme for transpositions unlinked to the original site. This was important as insertions at the original site were expressed, obscuring new patterns. Also important was Hong Ma's contribution. *Arabidopsis* transformation was very difficult in those days and Hong Ma, recently appointed to the CSH Laboratory staff from Eliot Meyerowitz's lab at the California Institute of Technology, generated all the original transformants. After the first few crosses, it was clear that the system worked well, surpassing our expectations.

Several discoveries quickly followed—the first gene expressed and required in the haploid gametophyte, *PROLIFERA* (replication licensing factor), was found by postdoc Patty Springer, who mutagenized the first few thousand lines.[1,2] The first imprinted gene *MEDEA* (Enhancer of Zeste) was found among these lines by Ueli Grossniklaus,[3] who joined us from Walter Gehrings' lab as a CSHL fellow. MEDEA controls seed size and development epigenetically. We sent the "starter" lines to colleagues around the world, who have since used the system to identify, for example, genes involved in disease resistance (a MAP kinase)[4] and root evolution (*root hair defective* 6).[5] Examples of commercially important findings were of transcription factors expressed in the fruit wall that control seed loss at harvest time, including *FRUITFULL, INDEHISCENT, ALCATRAZ,* and *REPLUMLESS*[6–9] (or *BELLRINGER,* named after the BELL transcription factor family[10]) all of which were functionally identified as gene-trap insertions.

As Jim had predicted, agricultural biotechnology companies were quickly interested in our growing collection of lines, which we mapped to the genome by a PCR (polymerase chain reaction) technique invented by Whittier and colleagues from Japan.[11] Along with Dick McCombie, we generated a database of insertions in a large fraction of *Arabidopsis* genes.[12] This resulted in funding from five multinational "agbiotech" companies, although allowing access to the same database at the same time was challenging. But the companies were very forgiving, and their support for the whole plant group continued for more than six years. Grant support from federal government agencies proved largely elusive, although colleagues did secure funding from the European Union and from Singapore. But the success of gene traps inspired a similar collection of T-DNA (as opposed to transposon) insertions from Joe Ecker (SALK Institute) and groups in France, Holland, and Germany, who between them have made *Arabidopsis* one of the few model organisms with close to a complete collection of gene "knockouts."[13]

The database of gene and enhancer traps has stood the test of time. We continue to receive orders for strains from around the world, and the CSHL collection continues to grow. The authors have gone on to bigger things. Sundar moved on first to Singapore and later to the University of California at Davis as Chair of Plant Biology. Ueli is now Professor of Botany in Zurich. Patty Springer is tenured at the University of California in Riverside, and Sam Haward, an undergraduate research program student (URP), went on to a Ph.D. with John Gray in Cambridge. Tom Volpe was an undergraduate volunteer, who later did his Ph.D. in yeast genetics, returning to my lab as a postdoc to make the *Science* "Breakthrough of the Year" in RNAi (RNA interference) almost 10 years later. He is now at Northwestern University in Chicago.

What, I wonder, would Barbara McClintock make of it all if she were to sit in on the seminars my graduate students give today, discussing not only transposition, but RNAi, too? I think that she would be thrilled. After all, her hope in the quotation at the head of this essay—If anyone wants a locus to be mutable, just put in the order and one will be sent the following year!—has come to pass, although the timescale wasn't quite right! But for her, transposition was not the thing. Rather, it was the ways in which transposons control gene expression. It is now clear that RNAi provides this exquisite specificity and may underlie many of the mysteries that made her work such an inspiration to us all.

Notes and References

1. Sundaresan V., Springer P., Volpe T., Haward S., Jones J.D.G., Dean C., Ma H., and Martienssen R. 1995. Patterns of gene action in plant development revealed by enhancer trap and gene trap transposable elements. *Genes Dev.* **9:** 1797–1810.

2. Springer P.S., McCombie W.R., Sundaresan V., and Martienssen R.A. 1995. Gene trap tagging of PROLIFERA, an essential MCM2-3-5-like gene in *Arabidopsis*. *Science* **268:** 877–880.

3. Grossniklaus U., Vielle-Calzada J.P., Hoeppner M.A., and Gagliano W.B. 1998. Maternal control of embryogenesis by MEDEA, a polycomb group gene in *Arabidopsis*. *Science* **280:** 446–450.

4. Petersen M., et al. 2000. *Arabidopsis* map kinase 4 negatively regulates systemic acquired resistance. *Cell* **103:** 1111–1120.

5. Menand B., Yi K., Jouannic S., Hoffmann L., Ryan E., Linstead P., Schaefer D.G., and Dolan L. 2007. An ancient mechanism controls the development of cells with a rooting function in land plants. *Science* **316:** 1477–1480.

6. Liljegren S.J., Roeder A.H.K, Kempin S.A., Gremski K., Østergaard L., Guimil S., Reyes D.K., and Yanofsky M.F. 2004. Control of fruit patterning in *Arabidopsis* by INDEHISCENT. *Cell* **116:** 843–853.

7. Gu Q., Ferrandiz C., Yanofsky M.F., and Martienssen R. 1998. The FRUITFULL MADS-box gene mediates cell differentiation during *Arabidopsis* fruit development. *Development* **125:** 1509–1517.

8. Rajani S. and Sundaresan V. 2001. The *Arabidopsis* myc/bHLH gene ALCATRAZ enables cell separation in fruit dehiscence. *Curr. Biol.* **11:** 1914–1922.

9. Roeder A.H., Ferrandiz C., and Yanofsky M.F. 2003. The role of the REPLUMLESS homeodomain protein in patterning the *Arabidopsis* fruit. *Curr. Biol.* **13:** 1630–1635.

10. Byrne M.E., Groover A.T., Fontana J.R., and Martienssen R.A. 2003. Phyllotactic pattern and stem cell fate are determined by the *Arabidopsis* homeobox gene BELL-RINGER. *Development* **130:** 3941–3950.

11. Liu Y.G., Mitsukawa N., Oosumi T., and Whittier R.F. 1995. Efficient isolation and mapping of *Arabidopsis thaliana* T-DNA insert junctions by thermal asymmetric interlaced PCR. *Plant J.* **8:** 457–463.

12. Martienssen R.A. 1998. Functional genomics: Probing plant gene function and expression with transposons. *Proc. Natl. Acad. Sci.* **95:** 2021–2026.

13. Alonso J.M., et al. 2003. Genome-wide insertional mutagenesis of *Arabidopsis thaliana. Science* **301:** 653–657.

Patterns of gene action in plant development revealed by enhancer trap and gene trap transposable elements

Venkatesan Sundaresan,[1,4] Patricia Springer,[1] Thomas Volpe,[1] Samuel Haward,[1,3] Jonathan D.G. Jones,[2] Caroline Dean,[2] Hong Ma,[1] and Robert Martienssen[1]

[1]Cold Spring Harbor Laboratory, Cold Spring Harbor, New York 11724 USA; [2]John Innes Centre, Norwich, UK

The crucifer *Arabidopsis thaliana* has been used widely as a model organism for the study of plant development. We describe here the development of an efficient insertional mutagenesis system in *Arabidopsis* that permits identification of genes by their patterns of expression during development. Transposable elements of the *Ac/Ds* system carrying the GUS reporter gene have been designed to act as enhancer traps or gene traps. A novel selection scheme maximizes recovery of unlinked transposition events. In this study 491 plants carrying independent transposon insertions were generated and screened for expression patterns. One-half of the enhancer trap insertions and one-quarter of the gene trap insertions displayed GUS expression in seedlings or flowers, including expression patterns specific to organs, tissues, cell types, or developmental stages. The patterns identify genes that act during organogenesis, pattern formation, or cell differentiation. Transposon insertion lines with specific GUS expression patterns provide valuable markers for studies of *Arabidopsis* development and identify new cell types or subtypes in plants. The diversity of gene expression patterns generated suggests that the identification and cloning of *Arabidopsis* genes expressed in any developmental process is feasible using this system.

[*Key Words*: *Arabidopsis* development; plant development; enhancer trap; gene trap; transposable elements]

Received April 14, 1995; revised version accepted June 15, 1995.

McClintock's investigations on alterations in the patterns of "gene action" in maize kernels led to the discovery of transposable elements (McClintock 1950). Subsequently, transposable elements have been found in almost all organisms examined and are believed to constitute a major agent for the generation of evolutionary diversity through mutations and genome rearrangements. The widespread distribution and mutagenic potential of these elements have led to their exploitation as valuable tools in genetic and molecular studies of prokaryotic and eukaryotic organisms. In particular, transposon mutagenesis has provided a means of cloning developmentally important genes from many higher eukaryotes, where cloning by complementation is not feasible. This successful utilization of transposon mutagenesis has led to the development of new approaches for gene identification and cloning, as described below.

In recent years the study of development in animals has been facilitated greatly by technologies using engineered insertion elements called enhancer traps and gene traps (for review, see Skarnes 1990; Wilson et al. 1990). The enhancer trap or gene trap element carries a reporter

gene construct that can respond to *cis*-acting transcriptional signals at the site of insertion. These elements permit the identification of genes by their pattern of expression and their subsequent cloning using the inserted element as a tag. A particularly useful aspect of this technology is that it permits the identification of genes that would have been missed in conventional mutagenesis screens. Many genes have multiple roles, acting at early as well as late stages in development, so that their later functions are obscured in phenotypic screens but are revealed in screens for gene expression patterns (Mlodzik et al. 1990; Wilson et al. 1990). Second, if a gene is functionally redundant because of the presence of a second locus that can subsitute for the same function, inactivation of the gene will not result in any phenotype. In the yeast *Saccharomyces cerevisiae*, it is estimated that 60%–70% of the genes have no phenotype upon disruption (Goebl and Petes 1986; Oliver et al. 1992; Burns et al. 1994). Such genes cannot be identified in screens for mutant phenotypes but may be detected by expression pattern in enhancer trap or gene trap screens.

In the last decade, the crucifer *Arabidopsis thaliana* has been used widely as a model system for studying plant development. Maize transposons can be made to transpose at sufficiently high rates in *Arabidopsis* to permit gene tagging and cloning (Aarts et al. 1993; Bancroft

[3]Present address: Department of Plant Sciences, Cambridge University, Cambridge, UK.
[4]Corresponding author.

Draetta G., Luca F., Westendorf J., Brizuela L., Ruderman J., and Beach D. 1989. **cdc2 protein kinase is complexed with both cyclin A and B: Evidence for proteolytic inactivation of MPF.** *Cell* **56:** 829–838. (Reprinted, with permission, from Elsevier ©1989.)

I DO NOT NORMALLY SUFFER FROM THE DISEASE of reading my old papers, though, occasionally, I vainly (in both senses) encourage new postdocs and students to do so. But when kindly asked to contribute to this volume I had no doubt about which piece from my Cold Spring Harbor days to choose, and having re-read the piece, the excitement of those days came flooding back. Thus, I struggle to be dispassionate. But let me start with my first big "science high," which although it occurred in a dark room (with no chemicals more intoxicating than the smell of photo-fixative) at the University of Sussex a few years earlier, now seems to lead inevitably toward the synthesis that was realized in the paper by Draetta (alias Giulio) et al. (1989).

In 1980, Paul Nurse and I had developed the methods for genetic transformation of fission yeast and thus were in the delightful position to play the game of seeking molecular relationships between the numerous cell cycle mutations of budding yeast and fission yeast, the function of virtually none of which were known.[1] I introduced a good library of budding yeast genes, made by Kim Nasmyth,[2] into a temperature-sensitive mutant of the *cdc2* gene of fission yeast mutant and rescued a gene. We had chosen *cdc2* because it had a rate-limiting role in G_2 and also appeared to function in G_1 as a "start" gene: It was special. So, which was the budding yeast *cdc2* homolog, if any? Steve Reed[3] had kindly provided gene isolates of all the budding yeast cell cycle "start" mutants that Hartwell and colleagues had identified,[3] and I did a "plasmid–plasmid" blot (don't ask why we didn't just sequence; it was early days), and it turned out that the gene that rescued *cdc2* was

DAVID BEACH came to Cold Spring Harbor in 1982. He left the Laboratory in 1998 to become President of Genetica, a biotechnology company that aims to integrate genetic data with biological function. David has held positions in London, at the Institute for Child Health, and more recently at the Institute for Cell and Molecular Science at Queen Mary's School of Medicine and Dentistry, where he is Professor of Stem Cell Biology. dhbeach@btinternet.com

CDC28. Thus, we had not only spanned, say, 1.5 billion years of evolution, about the same genetic time as humans and either of the yeast types, but also showed that the two "hottest" cell cycle mutants of the day were functionally equivalent. I stayed in the orange-light darkroom pondering the blot for quite some time. Having persuaded myself that there was no hideous artifact, I simply enjoyed the implications for biology and let's face it, my career; not immediately rushing out to tell Paul because he was a noisy fellow and would not only distract me, but also promptly tell the world (to be fair, the cell cycle field was not remotely competitive in those days and no one else was technically poised to get the result).

Fast forward to the summer of 1988 and the fledgling synthesis described above had progressed exceedingly well. *Cdc2/CDC28* had been found in all eukaryotes analyzed with a decent reagent. The previously mysterious factor that promoted mitosis (maturation promoting factor, MPF) by serial microinjection of mitotic cytoplasm into interphase amphibian eggs had been shown[4] by Giulio, myself, and John Newport's group (at the same time as the Maller and Nurse groups) to contain the cdc2/CDC28 protein. The same turned out to be the case for a very potent histone H1 kinase that was massively activated at mitosis and died thereafter. But the abundance of *cdc2* was stubbornly invariant during the cell cycle, in every species looked at.

Enter cyclins. These proteins, initially A and B on a gel, had been discovered in clam eggs by Joan Ruderman and Tim Hunt,[5] working in the summer course at Woods Hole, where I had once written my very poor Ph.D. thesis, while also taking a course. Fertilized marine invertebrate eggs have vast stores of just about everything and so do not make much except DNA. However, cyclins A and B oscillated in the early meiotic and mitotic cleavage divisions, due to very heavy synthesis and periodic degradation. Why? Lots of proteins were known to oscillate in a cell-cycle-dependent manner, but at least in eggs, these proteins were virtually the only proteins being synthesized after fertilization, hence their ready detection.

My first graduate student Bob Booher at Cold Spring Harbor had shown,[6] using classical fission yeast genetics (my second choice for Cold Spring Harbor Laboratory publication pride), a very close interaction between *cdc2* and *cdc13*, a gene we then isolated and, after a slight hiccup (a future Nobel Prize winner was responsible for homology searching at Cold Spring Harbor at the time), found to be a cyclin homolog. We hypothesized about biochemistry (as geneticists are wont to do) and suggested that *cdc13* might be either a subunit or a substrate of *cdc2* (both turned out to be true, the former the key issue). The stage was set. In the spring of 1988, Joan mentioned to me that anticyclin immunoprecipitates showed a weak kinase activity; she had also previously published that cyclin cDNA injection could activate frog eggs. The penny finally dropped. I suggested that she focus only on histone H1 as a substrate (positive) and so, particularly with Giulio's fantastic biochemical skills (he had previously shown that *cdc2* in yeast and humans was essentially the same protein with cleavage mapping; you won't guess which cleaving reagent he used), off we went. In brief, we showed that cyclins act as essential regulatory subunits of *cdc2*. My main fear was that the summer clam season would end (can you imagine my yeast incubators going down for a winter) before the work was submitted (or worse while we were fooling with reviewers' comments), but Joan assured me that she had ample frozen stocks of all clammy stuff.

We submitted to *Cell* in October 1988 and published in January 1989, but not before I nearly had a heart attack. On December 23 1988, Ben Lewin[7] gave me a special Christmas present. He published a minireview showing a key aspect of our work (a nice round cyclin molecule sticking to a nice round Cdc2 kinase; see the mini and the final figure of our paper) while our work was under revision in his journal. The citation for this conclusion was Bob Booher's genetics, a methodology that cannot resolve the issue. The 'phone line to Boston became incandescent. But oh it was fun to argue about other than whether my manuscript was suitable, in principle of course, for *Cell*. It already demonstrably was.

David Beach, 1985. *(Photograph by Susan Lauter, courtesy of Cold Spring Harbor Laboratory Archives.)*

That amusement aside, back to my first tale. After the publication of the *Cell* piece, we knew the following (apologies to all participants for ignoring the regulation of *cdc2* by phosphorylation here): *cdc2* was *CDC28*, and everyone, including us, seemed to have a version of the gene. *cdc2* was the kinase catalytic subunit of an entity called MPF that had direct mitosis-inducing powers, and histone H1 was a great substrate of the kinase, allowing us to assay M-phase activation and very abrupt inactivation of the kinase, despite no physical loss of the *cdc* subunit. Cyclins had been floating around, oscillating beautifully, and now, finally, cyclins acted as subunits of *cdc2* (subsequently to be renamed Cdk1, cyclin-dependent kinase). Equally importantly, cyclins A and B, which had rather different properties seemed to create separate Cdc2-bound kinases, thus allowing multiple roles for *cdc2*. Full dawning of the significance of this would await the discovery of G_1-acting budding yeast *cln* genes, human cyclin D, and all that.

To myself, a young Turk at the time (hopefully still so, at least in mind), the work was a grand synthesis of fungal cell cycle genetics (*cdc2/CDC28/cdc13*, etc), biochemistry in human cells and marine invertebrates (histone H1 M-phase kinase), and cell biology with elaboration of the MPF phenomenology and cyclin oscillations. For the first time, we seemed to have a genetic/cell biological/biochemical/molecular entity that incorporated the best of the past (incidentally demonstrating, once and for all, that no methodological approach was the sole way forward, at least in this field) and offering a platform for a new description of cell cycle control. I was rash enough to say so at conferences without noticeably winning friends in the field (could it have been the way I said it?) but was gloriously congratulated at a Chatham Bar Inn meeting in the fall of 1988 by Hal Weintraub[8] (may God care for the soul of that gentle, generous, and true scientist).

And now, dear reader, if you find all of this a little too subjective for your usual scientific rigor, see if you feel up to reading the paper, which was our best effort at objectivity in the glorious summer of 1988. And thank you Guilio, Leo, Frank, Joanne, and, of course, Joan herself.

Notes and References

1. Beach D. and Nurse P. 1981. High-frequency transformation of the fission yeast *Schizosaccharomyces pombe*. *Nature* **290**: 140–142. The work for this paper was carried out at the University of Sussex, where Paul Nurse was a faculty member. Nurse later became the Director-General of the Imperial Cancer Research Fund at Lincoln's Inn Fields, London, and then President of The Rockefeller University, New York.

2. Nasmyth K.A. and Tatchell K. 1980. The structure of transposable yeast mating type loci. *Cell* **19**: 753–764.

3. Hartwell L.H. and Weinert T.A. 1989. Checkpoints: Controls that ensure the order of cell cycle events. *Science* **246**: 629–634.

   Reed S.I. 1980. The selection of *S. cerevisiae* mutants defective in the start event of cell division. *Genetics* **95**: 561–577.

   Ferguson J., Ho J.Y., Peterson T.A., and Reed S.I. 1986. Nucleotide sequence of the yeast cell division cycle start genes CDC28, CDC36, CDC37, and CDC39, and a structural analysis of the predicted products. *Nucleic Acids Res.* **14**: 6681–6697.

4. Draetta G., Luca F., Westendorf J., Brizuela L., Ruderman J., and Beach D. 1989. Cdc2 protein kinase is complexed with both cyclin A and B: Evidence for proteolytic inactivation of MPF. *Cell* **56**: 829–838.

5. Rosenthal E.T., Hunt T., and Ruderman J.V. 1980. Selective translation of mRNA controls the pattern of protein synthesis during early development of the surf clam, *Spisula solidissima*. *Cell* **20**: 487–494.

6. Booher R. and Beach D. 1987. Interaction between cdc13+ and cdc2+ in the control of mitosis in fission yeast; dissociation of the G_1 and G_2 roles of the cdc2+ protein kinase. *EMBO J.* **6**: 3441–3447.

7. Ben Lewin was the founder and editor of *Cell*, a journal that revolutionized both the style and substance of scientific publishing when its glossy pages, spacy layout, and lovely illustrations first appeared in 1974. Lewin assured himself of a supply of excellent papers for the first few volumes by promising that any manuscripts from members of his original editorial board would be accepted for publication without demur.

8. Harold Weintraub worked at the Fred Hutchinson Cancer Research Center, Seattle, from 1978 until his death from cancer in 1995 at the age of 49. He solved scientific problems with a mixture of strong intuition, astute intelligence, and infectious charm. Among his best work was the identification of *myoD*, a gene that can elicit the entire program of muscle differentiation when introduced into nonmuscle cell types.

Cell, Vol. 56, 829–838, March 10, 1989, Copyright © 1989 by Cell Press

cdc2 Protein Kinase Is Complexed with Both Cyclin A and B: Evidence for Proteolytic Inactivation of MPF

Giulio Draetta,* Frank Luca,† Joanne Westendorf,†
Leonardo Brizuela,* Joan Ruderman,†
and David Beach*
* Cold Spring Harbor Laboratory
Cold Spring Harbor, New York 11724
† Department of Zoology
Duke University
Durham, North Carolina 27706

Summary

In the clam, Spisula, two previously described proteins known as cyclin A and B display the unusual property of selective proteolytic degradation at the end of each mitosis. We show here that clam oocytes and embryos contain a cdc2 protein kinase. This protein kinase is a component of the M phase promoting factor (MPF) in frog eggs and the M phase–specific histone H1 kinase in starfish. Clam cdc2 is found in association with both cyclin A and B, probably not as a trimolecular association, but as separate cdc2/cyclin A and cdc2/cyclin B complexes. Clam cdc2 and the associated cyclins bind to p13^{suc1}-Sepharose. The p13-bound complex, and also anti-cyclin A or B immunoprecipitates, each display cell cycle–dependent histone H1 kinase activity. We suggest that in addition to the cdc2 protein kinase, the cyclins are further components of the M phase promoting factor and that cyclin proteolysis provides the mechanism of MPF inactivation and thus exit from mitosis.

Introduction

Mitotic cyclins were first identified in embryos of marine invertebrates but are now known to exist in a wide range of, and possibly all, dividing eukaryotic cells (Evans et al., 1983; Swenson et al., 1986; Standart et al., 1987; Pines and Hunt, 1987; Booher and Beach, 1988; Solomon et al., 1988; Goebl and Byers, 1988). They were initially recognized on the basis of two distinctive properties. First, they are among the most abundantly synthesized polypeptides in activated oocytes and early embryos and are therefore readily detectable among whole cell proteins resolved by gel electrophoresis (see Figure 3, lane 1, also Rosenthal et al., 1980, 1982, 1983; Evans et al., 1983). Second, these proteins accumulate during interphase, but undergo abrupt degradation at the end of each mitosis. The resulting oscillation in cyclin levels is regulated by selective proteolysis, which occurs at the metaphase/anaphase transition (Evans et al., 1983; Swenson et al., 1986; Standart et al., 1987). In clams, there are two such proteins, known as cyclin A and B. They share 31% amino acid identity but are readily distinguished by differences in the precise pattern of accumulation and destruction during meiosis and mitosis (Swenson et al., 1986; Westendorf et al., submitted).

Several observations indicate that cyclin synthesis and degradation might play a role in mitosis. Firstly, overall protein synthesis is essential for passage from interphase into M phase, even in embryonic cells that have large stores of many of the proteins required for the cell division cycle (Wasserman and Masui, 1975; Newport and Kirschner, 1984; Gerhart et al., 1984; Meijer and Pondaven, 1988). More critically, introduction into Xenopus oocytes of mRNA that directs synthesis of either clam cyclin A or B (Swenson et al., 1986; Westendorf et al., submitted), or a sea urchin cyclin (Pines and Hunt, 1987), induces maturation. Oocyte maturation is associated with release of the cell from G2/prophase arrest and progression into meiotic M phase.

Entry of oocytes or somatic cells into M phase is induced by an activity known as MPF (maturation or M phase promoting factor, Smith and Ecker, 1971; Masui and Markert, 1971). MPF is a nonspecies-specific mitotic inducer (Kishimoto and Kanatani, 1976; Sunkara et al., 1979; Kishimoto et al., 1984; Tachibana et al., 1987) that directly triggers entry of an interphase cell into mitosis without requirement for further protein synthesis (Wasserman and Masui, 1975; Gerhart et al., 1984). MPF is usually detected by microinjection of the factor into frog or starfish oocytes, but it can also be assayed in cell-free lysates (Miake-Lye and Kirschner, 1985; Lohka and Maller, 1985; Dunphy and Newport, 1988). Active MPF can only be found in cells that are in or entering M phase (Wasserman and Smith, 1978; Doree et al., 1983; Gerhart et al, 1984), but various experiments have indicated that the mitotic inducer might exist in an inactive state during interphase. Under certain conditions, it can spontaneously reactivate in vitro without apparent requirement for protein synthesis (Cyert and Kirschner, 1988; Dunphy and Newport, 1988). It has been proposed that mitotic cyclins might function either directly or indirectly as activators of MPF (Evans et al., 1983; Swenson et al., 1986; Standart et al., 1987).

Recently, it has been established that one component of Xenopus MPF, and also of the starfish M phase–specific histone H1 kinase, is a homolog of the fission yeast cdc2 protein kinase (Dunphy et al., 1988; Gautier et al., 1988; Labbe et al., 1988; Arion et al., 1988). This kinase is also related to the product of the budding yeast *CDC28* cell cycle "start" gene (Hartwell, 1974; Beach et al., 1982). In fission yeast, *cdc2* regulates entry into mitosis (Nurse and Thuriaux, 1980) and in human tissue culture cells the equivalent 34 kd protein kinase (Lee and Nurse, 1987; Draetta et al., 1987) is activated at least 70-fold as cells move from G1 into mitosis (Draetta and Beach, 1988; Draetta et al., 1988b). The enzyme is maximally active during metaphase but is abruptly inactivated during the completion of cell division. Inactivation is associated with loss of a 62 kd subunit from the protein kinase complex (Draetta and Beach, 1988). The properties of this 62 kd protein are consistent with it being a mitotic cyclin, but it is presently not known whether this protein is actually degraded during mitosis or simply dissociates from the cdc2 complex for other reasons.

DNA

Why the Revolution That Jim Started Continues

BRUCE ALBERTS

———

P ERHAPS THE MOST FAMOUS ONE-LINER IN THE HISTORY of science is the comment that "It has not escaped our notice that the specific pairing we have postulated immediately suggests a possible copying mechanism for the genetic material," with which Jim Watson and Francis Crick ended their famous paper (Watson and Crick 1953).[1] I began my life as a scientist in 1959 as a Harvard undergraduate in the laboratory of Paul Doty, and I remained there until 1965 as a graduate student. I interacted with Jim frequently at Harvard, as he made himself remarkably accessible to students. But even for those not in close proximity to Jim, it was impossible not to be inspired by his great discovery.

To some, the purification and characterization of DNA polymerase by Arthur Kornberg (Kornberg 1960)[2] seemed to complete the DNA replication story. But DNA polymerase requires single-stranded DNA as a template, and the DNA in a cell is double-stranded. In graduate school, I tried, quite unsuccessfully, to solve this dilemma (Alberts 2004).[3] Only in 1965, after I had arrived in Geneva, Switzerland, for a year of postdoctoral work with Alfred Tissieres, did I realize how igno-

———

BRUCE ALBERTS received his Ph.D. from Harvard University in 1965 and is now professor of Biochemistry and Biophysics at the University of California, San Francisco, where he moved after ten years at Princeton. In addition to his research, he has also pursued interests in science education and policy, serving as president of the U.S. National Academy of Sciences and chair of the National Research Council. He is one of the original authors of the influential textbook *The Molecular Biology of the Cell* and was recently appointed the 18th Editor-in-Chief of the journal *Science*.

31

rant I had been in graduate school. Much of my thesis work had been predicated on the assumption that only a single protein (the DNA polymerase) was needed for DNA replication. Had I attended the 1963 Cold Spring Harbor Symposium, I would have certainly known of the elegant genetic analyses of Robert Edgar, Richard Epstein, and their collaborators at the California Institute of Technology, who had shown that at least seven viral gene products were required for the T4 bacteriophage to replicate its DNA (Epstein et al. 1963).[4] My sudden recognition of this fact was quite exciting, because there were not even guesses for what the six extra proteins (those besides the polymerase) might be doing in the replication process.

There was also the puzzle of how a single type of DNA polymerase that adds nucleotides only in the 5′ to 3′ direction could copy both strands of the double helix. As Watson and Crick had found, these strands are paired in opposite polarity. Thus, I distinctly remember a professor at Harvard suggesting to us in 1960 that each unit of DNA in a chromosome might be replicated by two DNA polymerase molecules, one starting from each end of a long double helix. In this case, the moving polymerase molecule would displace one of the DNA strands and leave it unpaired as an enormously long single strand. After the two DNA polymerase molecules passed each other near the center of the molecule, this single-stranded DNA would be used as a template by the polymerase molecule heading in the opposite direction, finally converting it to a double helix.

The discovery of the replication fork by John Cairns, announced at the 1961 Cold Spring Harbor Symposium (Cairns 1961),[5] put an end to such speculation, but it left many mysteries—as symbolized by drawing the fork covered by a fig leaf in the 1960s. It was only after Reiji and Tomoko Okazaki observed that newly synthesized DNA consists of a preponderance of very short DNA fragments (Okazaki et al. 1968)[6] that the idea of discontinuous DNA synthesis became accepted, with one strand at the fork (the lagging strand) being made by a back-stitching mechanism while the strand that grows in the 5′ to 3′ direction is made continuously (the leading strand).

This brings us to the first paper in this section, in which Jim Watson recognizes that the mechanism of DNA synthesis on the lagging strand creates a serious problem for completing the replication of each end of a double helix (Watson 1972; see also Sambrook, p. 37).[7] This recognition has produced a growth industry, in part because the solution to this problem in eukaryotic cells (telomeres, specially replicated by a telomerase enzyme) represents a major target for anticancer drug development. (A current search of the literature for "telomere" produces a list of more than 9000 publications, 1500 of which are review articles.) The existence of special DNA sequences at telomeres is also a reminder that, even though each eukaryotic chromosome consists of a single enormously long DNA molecule, it is punctuated by special structures—not only its two telomeres, but also a centromere and many replication origins.

Biochemists soon began to discover that proteins can do many quite remarkable things on DNA. Some of these are so simple and elegant that one wonders why they were not predicted before experiments revealed them. The ring-shaped structure of the sliding clamp that is used to keep a single DNA polymerase moving processively along the same DNA template strand is one striking example. As reported in the second paper in this section, the eukaryotic version of this clamp, proliferating cell nuclear antigen (PCNA), was discovered in Bruce Stillman's laboratory (Prelich et al. 1987; see also Stillman, p. 41).[8] Four years later, Michael O'Donnell and his collaborators would demonstrate that its bacterial homolog forms a ring around the DNA helix that keeps the DNA polymerase tethered to the same molecule as it moves (Stukenberg et al. 1991; Kong et al. 1992).[9,10] My laboratory had discovered the first sliding clamp as part of the T4 bacteriophage replication system (Huang et al. 1981; Mace and Alberts 1984),[11,12] and we spent many years probing its mechanism without great success. In retrospect, it is hard to comprehend why I never imagined the "ring as a tether" possibility until O'Donnell's work. This is one of an enormous number of examples that I could cite to support the conclusion that "cells are much smarter than scientists."

Both PCNA and the bacterial rings are loaded onto the DNA at a replication fork by a special clamp-loader protein complex. After their release from the DNA polymerase, the rings can locally diffuse rapidly back and forth along the DNA helix, thereby serving as a mark for a nearby replication fork (Tinker et al. 1994).[13] This fact is exploited to attract many other proteins to regions of newly synthesized DNA. Amazingly, a recent review lists more than 50 proteins that bind to PCNA (Moldovan et al. 2007).[14] Each is thought to bind to PCNA in a highly regulated way, so as to help catalyze such functions as DNA repair (bypass replication, base excision repair, nucleotide excision repair, mismatch repair), chromatin formation (histone disposition on newly replicated DNA, chromatin remodeling), sister chromatid cohesion (loading of cohesin), and the formation of localized "replication factories," each of which contains a cluster of a dozen or so replication forks.

The third paper in this section reports a major breakthrough in the DNA replication field: the identification of the protein complex that binds to DNA in order to specify each replication origin—the sites where DNA replication forks will be initiated during S phase (Bell and Stillman 1992; see also Stillman, p. 47).[15] In the yeast *Saccharomyces cerevisiae*, where this work was done, replication origins are specified by a set of specific DNA sequences, as they are in bacteria. But it turned out to be much more difficult to identify the proteins that recognize these origins than it had been in bacteria. Rather than a single protein, a protein complex was discovered that Bell and Stillman named ORC, for origin recognition complex.

A large amount of work has been done to locate the sites where DNA replication forks begin in mammalian cells, with very confusing results. Here, specific

DNA sequences appear to play a considerably smaller role in localizing an origin than do other features of the DNA, such as its packaging in chromatin (Cvetic and Walter 2005).[16] Nevertheless, a few year's later, Stillman's laboratory would use the yeast information to show that a very similar ORC complex is used to mark each replication origin in mammals (Gavin et al. 1995).[17]

The activity of an ORC complex is highly regulated during the cell cycle—first to make sure that no DNA is replicated until a cell makes the critical decision to proliferate and then to guarantee that every region of the DNA helix in the enormously long DNA molecule that forms a chromosome is replicated once, and only once, in each S phase (Diffley 2004).[18] Thus, the discovery of ORC marked a major advance in our understanding of cells—one that has increased our appreciation of the sophisticated regulatory mechanisms that serve to control the activity of this and many other DNA protein complexes.

The final paper in this section represents an advance in our understanding of the dynamic processes in DNA that allow its many different features to be specifically recognized by proteins (Klimašauskas et al. 1995; see also Roberts, p. 51).[19] Stimulated by this discovery from the Roberts laboratory, we now know that the DNA bases spontaneously flip out and in of the double helix on a millisecond timescale. However, this is not fast enough for some of the proteins that scan the DNA helix for specific features, and many enzymes (such as the HhaI methyltransferase and uracil DNA glycosylase) appear to catalyze a base-flipping reaction as they slide rapidly along double-stranded DNA molecules (Bouvier and Grubmuller 2007).[20]

The collection of papers in this volume demonstrates how the research carried out at the Cold Spring Harbor Laboratory has played, and is continuing to play, a central role in the rapid advance of the biological sciences. But there is an enormous amount more that needs to be done, especially as the chemistry of life is much more sophisticated than most of us imagined even ten years ago. We now recognize that nearly every biological process is catalyzed by a set of ten or more interacting proteins that undergo highly ordered movements in a machine-like assembly (Alberts 1984, 1998).[21,22] Moreover, these protein complexes frequently change both their activity and their locations inside the cell in response to intracellular signals.

The protein machines that act on DNA are highly regulated. Although the replication, repair, and recombination of the DNA double helix are often considered as separate, isolated processes, the same DNA molecule is able to undergo any one of these reactions inside the cell. Moreover, the same protein may participate with different protein partners in specific combinations of the three types of reactions. For instance, DNA recombination is often linked directly to either DNA replication or DNA repair. For the integrity of a chromosome to be properly maintained, each specific reaction must be carefully directed and controlled. This requires that sets of proteins be assembled on the DNA and activated only where and when they are needed.

Although much remains to be learned about how these choices are made, it seems that specialized proteins that serve as "assembly factors" recognize different types of DNA structures. These assembly factors are only activated when they are appropriately covalently modified, and the modifications reflect the state of the cell (e.g., the stage of the cell cycle, whether the DNA is damaged, or whether the cell cycle has temporarily stalled at a checkpoint as a protective measure). Each assembly factor then serves to nucleate the assembly of the particular set of proteins needed for catalyzing the reaction appropriate to that time and place in the cell.

In summary, the revolution that Jim Watson started in 1953 will continue for many decades to come. There is therefore little doubt that the unique Laboratory that he has invigorated with his boundless energy and creative insights will continue to be one of the world's most exciting centers for new scientific discoveries, throughout the 21st century.

Notes and References

I presented my first paper as an independent investigator at a Cold Spring Harbor Symposium: Alberts B.M., Amodio F.J., Jenkins M., Gutmann ED., and Ferris F.L. 1969. Studies with DNA-cellulose chromatography. I. DNA-binding proteins from *Escherichia coli. Cold Spring Harbor Symp. Quant. Biol* 33: 289–305.

1. Watson J.D. and Crick F.H.C. 1953. Molecular structure of nucleic acids: A structure for deoxyribose nucleic acid. *Nature* 171: 737–738.

2. Kornberg A. 1960. Biological synthesis of DNA. *Science* 131: 1503–1508.

3. Alberts B. 2004. A wake-up call: How failing a PhD led to a strategy for a successful scientific career. *Nature* 431: 1041.

4. Epstein R.H., Bolle A., Steinberg C.M., Kellenberger E., Edgar R.S., Susman M., Denhardt G.H., and Lielausis A. 1963. Physiological studies of conditional lethal mutants of bacteriophage T4D. *Cold Spring Harbor Symp. Quant. Biol.* 28: 375–392.

5. Cairns J. 1963. The bacterial chromosome and its manner of replication as seen by autoradiography. *J. Mol. Biol.* 6: 208–213.

6. Okazaki R., Okazaki T., Sakabe K., Sugimoto K., and Sugino A. 1968. Mechanism of DNA chain growth: Possible discontinuity and unusual secondary structure of newly synthesized chains. *Proc. Natl. Acad. Sci.* 59: 598–605.

7. Watson J.D. 1972. Origin of concatemeric T7 DNA. *Nat. New Biol.* 239: 197–201.

8. Prelich G., Kostura M., Marshak D.R., Mathews M.B., and Stillman B. 1987. The cell-cycle regulated proliferating cell nuclear antigen is required for SV40 DNA replication *in vitro. Nature* 326: 471–475.

9. Stukenberg P.T., Studwell-Vaughan P.S., and O'Donnell M. 1991. Mechanism of the sliding beta-clamp of DNA polymerase III holoenzyme. *J. Biol. Chem.* **266:** 11328–11334.

10. Kong X.P., Onrust R., O'Donnell M., and Kuriyan J. 1992. Three-dimensional structure of the beta subunit of *E. coli* DNA polymerase III holoenzyme: A sliding DNA clamp. *Cell* **69:** 425–437.

11. Huang C.C., Hearst J., and Alberts B. 1981. Two types of replication proteins increase the rate at which T4 DNA polymerase traverses the helical regions in a single-stranded DNA template. *J. Biol. Chem.* **256:** 4087–4094.

12. Mace D.C. and Alberts B.M. 1984. Characterization of the stimulatory effect of T4 gene 45 protein and the gene 44/62 protein complex on DNA synthesis by T4 DNA polymerase. *J. Mol. Biol.* **177:** 313–327.

13. Tinker R.L., Kassavetis G.A., and Geiduschek E.P. 1994. Detecting the ability of viral, bacterial and eukaryotic replication proteins to track along DNA. *EMBO J.* **13:** 5330–5337.

14. Moldovan G.L., Pfander B., and Jentsch S. 2007. PCNA, the maestro of the replication fork. *Cell* **129:** 665–679.

15. Bell S.P. and Stillman B. 1992. ATP-dependent recognition of eukaryotic origins of DNA replication by a multiprotein complex. *Nature* **357:** 128–134.

16. Cvetic C. and Walter J.C. 2005. Eucaryotic origins of replication: Could you please be more specific? *Semin. Cell Dev. Biol.* **16:** 343–353.

17. Gavin K.A., Hidaka M., and Stillman B. 1995. Conserved initiator proteins in eukaryotes. *Science* **270:** 1667–1671.

18. Diffley J.F.X. 2004. Regulation of early events in chromosome replication. *Curr. Biol.* **14:** R778–R786.

19. Klimašauskas S., Kumar S., Roberts R.J., and Cheng X. 1994. HhaI methyltransferase flips its target base out of the DNA helix. *Cell* **76:** 357–369.

20. Bouvier B. and Grubmuller H. 2007. A molecular dynamics study of slow base flipping in DNA using conformational flooding. *Biophys. J.* **93:** 770–786.

21. Alberts B. 1984. The DNA enzymology of protein machines. *Cold Spring Harbor Symp. Quant. Biol.* **49:** 1–12.

22. Alberts B. 1998. The cell as a collection of protein machines: Preparing the next generation of molecular biologists. *Cell* **92:** 291–294.

JOSEPH F. SAMBROOK

Watson J.D. 1972. **Origin of concatemeric T7 DNA.** *Nat. New Biol.* **239:** 197–201. (Reprinted, with permission, from Macmillan Publishers Ltd ©1972.)

I N HIS DIRECTOR'S REPORT FOR 1971, Jim Watson described how the newly constructed addition to James Laboratory allowed him "to have my office in a building where I can talk science, not administration." This paper was written in that office during the last weeks of summer of the following year. I remember going to talk to him about something or other one sunlit afternoon when he was working on the first draft—his left hand crabbing across the paper to form his miniscule, slanted script. It seems incongruous that prose of such clarity should emerge from these cramped cryptograms.

By the beginning of the 1970s, the conundrum of how the ends of linear DNAs were replicated had become acute.[1] All enzymes capable of synthesizing DNA were known to move in the same direction along a DNA template: All required a primer and none were capable of starting synthesis at the very end of a duplex DNA molecule. When these constraints were applied to the replication of linear DNA molecules, the inevitable result was loss of sequences from the 5′ termini of the daughter DNA strands, with each cycle of replication yielding progressively shorter progeny molecules.

Jim's solution to the problem was both elegant and simple. Using bacteriophage T7 as an example, he showed that the end-replication problem could simply be avoided if both ends of the linear template DNA contained the same sequence. The sticky-ended progeny molecules generated by replication of the template could then use Watson-Crick pairing to form linear dimers whose internal gaps could be filled by a standard repair reaction. This cycle of replication, pairing, and gap repair could

---

JOE SAMBROOK is currently Distinguished Fellow at the Peter MacCallum Cancer Centre, Melbourne, Australia and a Professor in the Department of Pathology at the University of Melbourne. Between 1969 and 1985, Sambrook worked at Cold Spring Harbor Laboratory in New York, before moving to the University of Texas Southwestern Medical School as Chairman of Biochemistry. He returned to Australia in 1995 to be Director of Research at the Peter MacCallum Cancer Centre, Melbourne. joe.sambrook@petermac.org

Jim Watson, circa 1968. *(Courtesy of Cold Spring Harbor Laboratory Archives.)*

then be repeated to generate large DNA concatemers from which unit-length progeny bacteriophage genomes could be recovered by cleavage and end-repair.

Jim rather cryptically suggests that this mechanism, which so satisfyingly explained the replication of bacteriophage T7 DNA, might also apply to other linear DNAs. However, it was soon clear that the enzymology of DNA replication alone could not account for the properties of all types of DNA ends—in particular, the telomeric regions of eukaryotic chromosomes. During the years following the publication of Jim's paper, a number of alternative schemes for the structure and function of telomeres were proposed,[2-4] each with an attractive aspect, but none completely persuasive. In the end, as Carol Greider describes on page 75 of this book, it was to be another decade before the complex structure and biological importance of telomeres was understood.

Notes and References

1. Olovnikov A.M. 1971. Principles of marginotomy in template synthesis of polynu-cleotides. *Dokl. Akad. Naut.* **201:** 1496–1499.

2. Olovnikov A.M. 1973. A theory of marginotomy. The incomplete copying of tem-plate margin in enzymic synthesis of polynucleotides and biological significance of the phenomenon. *J. Theor. Biol.* **41:** 181–190.

3. Bateman A.J. 1975. Simplification of palindromic telomere theory. *Nature* **253:** 379–380.

4. Cavalier-Smith T. 1974. Palindromic base sequences and replication of eukaryotic chromosome ends. *Nature* **250:** 467–470.

NATURE NEW BIOLOGY VOL. 239 OCTOBER 18 1972

Origin of Concatemeric T7 DNA

J. D. WATSON

The Biological Laboratories, Harvard University, and the Cold Spring Harbor Laboratory

The observed concatemers of T7 DNA are consistent with replication schemes resulting in double-helical molecules with 3′ ended tails. Right-ended and left-ended molecules can then join to form dimers which on further replication similarly form larger concatemers.

THE replication pattern of T7 DNA at first site appears disarmingly simple. Initiation begins at an internal site 17% from the left end with replication proceeding in both directions to give rise first to an eye-shaped intermediate containing a replicating bubble, then to a Y-shaped form, and finally to two unit length daughter molecules (Fig. 1)[1,2]. Given the simple way the original infecting molecules replicate, we might expect that subsequent cycles of DNA replication would also generate unit length DNA molecules. Later in infection, however, most newly synthesized T7 DNA is present in concatemers of length two, three, four times . . . that of the original parental molecule[3-5]. With time these concatemers break down to unit length molecules that become incorporated into mature virus particles.

Equally puzzling has been the fact that the two ends of the T7 molecule are genetically equivalent, containing the same sequence of about 260 base pairs at both ends of the molecule[6]. At first, this finding prompted the suggestion that circular intermediates could be generated by limited digestion with exonuclease. Attack of infecting T7 molecules by an enzyme like *Escherichia coli* exonuclease III would convert them into forms with sticky ends like those of phage λ ; these would be expected to circularize the moment they are formed. But when Wolfson, Dressler, and Magazin[1,2] found that the first cycle of T7 replication did not appear to involve a circular form, the significance of the redundant ends became again an open question.

I shall show that consideration of linear replication in terms of our knowledge of the enzymology of DNA replication leads to the expectation that the replication of linear molecules necessarily generates concatemers which are integral multiples of the mature viral chromosome. For this process to occur, terminally repetitious sequences are prerequisite.

Growth within Replication Bubbles

Electron microscopy of the replicating bubbles reveals three main states[7] (Fig. 2). In the first category (*a*) all the DNA within the bubble appears double-stranded, including the regions immediately adjacent to the growing forks. The second group (*b*) comprises bubbles with one single-stranded section always located next to the growing fork. In the third group (*c*) there are two single-stranded sections, one at each growing fork and always situated in *trans*. Failure to observe the two single-strand regions in *cis* orientation confirmed the prevailing belief that the two chain ends (5′ and 3′) at each replication fork must elongate in different ways, with spurts of 5′–3′ growth going before the 3′–5′ growth in the complementary strand. Delius *et al.*[8] have directly proved that the 5′ ends abut onto the single-stranded portions of replicating T4 DNA. All such single-

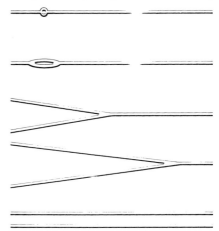

Fig. 1 Schematic representation of the first cycle of T7 DNA replication.

Prelich G., Kostura M., Marshak D.R., Mathews M.B., and Stillman B. 1987. **The cell-cycle regulated proliferating cell nuclear antigen is required for SV40 DNA replication** *in vitro*. *Nature* **326**: 471–475. (Reprinted, with permission, from Macmillan Publishers Ltd ©1987.)

T HE PAPER DISCUSSED HERE IS THE FIRST of a pair published back-to-back in *Nature* in 1987.[1,2]

In 1978, while I was still a graduate student,[3] Robert Tjian, Georg Fey, and Adolf Graessmann at Cold Spring Harbor reported[4] that simian virus 40 (SV40) large tumor antigen (T antigen) could stimulate cell DNA replication when microinjected into mammalian cells. The timing and course of chromosomal replication in mammalian cells are normally controlled with great precision. However, the results of Tjian et al. showed that purified T antigen, the protein product of an oncogene, could override the normal cellular controls and could promote the initiation of chromosomal DNA replication. These observations had obvious relevance both to the mechanism by which SV40 could transform mammalian cells and to the events that enable chromosomes to be replicated in normal cells.

When I came to Cold Spring Harbor Laboratory as a postdoctoral fellow in 1979, my goal was to characterize and codify the protein machinery that enabled chromosomes to be copied. An initial aim was to discover how DNA replication was turned on by an oncogene, a project that required T antigen in sufficient quantities for large-scale biochemistry. By 1983, this logistical problem had been solved by Yakov (Yasha) Gluzman, who had constructed a set of recombinant adenoviruses that synthesized useful amounts of SV40 T antigen.[5] In late 1984, Thomas Kelly and his then graduate student Joachim Li reported[6] that extracts of SV40-infected cells could support the replication of plasmid DNAs containing the SV40 origin of DNA replication (*ori*). Soon afterward, they[7] and Gluzman and I[5] reported that extracts from human cells sup-

---

BRUCE STILLMAN, a native of Australia, moved to Cold Spring Harbor Laboratory as a postdoctoral fellow in 1979 and has been at the Laboratory ever since, being promoted to the scientific staff in 1981. In 1994, he became the Director of the Laboratory and is currently its President. stillman@cshl.edu

Bruce Stillman, Terri Grodzicker, and Yasha Gluzman, 1983. *(Photograph by Joan James, courtesy of Cold Spring Harbor Laboratory Archives.)*

ported SV40 *ori-*dependent replication in the presence of recombinant T antigen. I also noticed that inclusion of a protein fraction prepared from the nucleus of human cells caused negative supercoiling of the replicated DNA due to replication-dependent chromatin assembly. The stage was set to identify cellular proteins present in the extract, a feat achieved by biochemical fractionation of the essential proteins.

Greg Prelich, my first graduate student, worked on adenovirus DNA replication for his rotation project, producing a good paper.[8] When he returned to my lab as a full-time student, he set out to fractionate proteins extracted from a line of human cells (293) that supported efficient replication in vivo of SV40 *ori* plasmids in the presence of T antigen. By then, I had used ion-exchange chromatography to obtain three protein fractions, each of which was required for replication. After further fractionation by Greg, one of these preparations yielded a fraction that stimulated DNA synthesis but, interestingly, was not essential for initiation of DNA replication. For his Ph.D. project, Greg proposed to purify the active protein(s), but his thesis committee rejected his proposal because they were not convinced that a fraction that was not required for initiation of DNA synthesis would yield much of interest, particularly since we were unsure how many proteins were in the fraction.

Dejected, I suggested that Greg continue with the work and within a very short time, he had purified a 36-kD protein as the active and only component. This was the first cellular protein to be identified using the SV40 DNA replication system. Greg demonstrated that the protein stimulated DNA replication and collaborated with John Smart[9] to sequence the protein by Edman degradation. A most pleasant surprise came when we learned that the protein was identical to the proliferating cell nuclear antigen (PCNA), a protein with an alterative name of cyclin (not the same as the now famous cyclin discovered previously by Tim Hunt in 1983[10]). This coincidence was the more amazing because Mike Mathews,[11] cohead of the next-door laboratory, where I had worked for my first three years at Cold Spring Harbor, was studying human autoantigens, one of which was PCNA.

PCNA/cyclin had been originally identified by Eng Tan,[12] a rheumatologist at the Scripps Clinic, as an antigen recognized by some patients with systemic lupus erythromatosis, an autoimmune disorder. Mathews and his postdoctoral fellow Matt Kostura had demonstrated in 1984[13] that PCNA was equivalent to cyclin, a protein identified by two-dimensional gel electrophoresis to vary in abundance during the cell division cycle. PCNA/cyclin associated in a dynamic pattern with chromatin, becoming bound during the DNA synthesis phase of the cell cycle and separating from chromatin at other times.

The observation that PCNA was required for DNA replication from the SV40 *ori* was exciting; but later, others suggested that PCNA was no more than an inhibitor of a protein that itself inhibited DNA replication, hence explaining why it was not essential. We noticed, however, in a paper published from Antero So and Kathleen Downey in late 1986[14] that a 36-kD protein could stimulate the activity of a then little-known DNA polymerase called DNA polymerase δ. I contacted So and Downey and soon thereafter we demonstrated first that PCNA was identical to their protein and second that PCNA stimulated DNA synthesis by increasing the processivity of polymerase δ, causing the polymerase to synthesize longer strands of DNA. In early 1987, two papers were submitted to *Nature*, one[1] reporting the initial partition of the extract of 293 cells into multiple fractions and the identification of PCNA as a new protein involved in DNA replication. The second paper[2] reported the stimulation of polymerase δ processivity. Subsequent studies demonstrated that PCNA was a sliding clamp that recruits many proteins to the DNA replication fork, including proteins that assemble chromatin.

PCNA is now one of the most used diagnostic and prognostic markers for human cancer, accurately reporting the percentage of S-phase cells in tissue. Most importantly, at least to Greg Prelich, was taking his new results to his baulky thesis committee and getting a passing grade.

Notes and References

1. Prelich G., Kostura M., Marshak D.R., Mathews M.B., and Stillman B. 1987. The cell-cycle regulated proliferating cell nuclear antigen is required for SV40 DNA replication *in vitro*. *Nature* **326**: 471–475.

2. Prelich G., Tan C.K., Kostura M., Mathews M.B., So A.G., Downey K.M., and Stillman B. 1987. Functional identity of proliferating cell nuclear antigen and a DNA polymerase-δ auxiliary protein. *Nature* **326:** 517–520.

3. At the Australian National University, Canberra, where his supervisor was Dr. Alan Bellett. His Ph.D. thesis "Mechanism of eukaryotic DNA replication" is an analysis of the steps involved in the initiation of replication of the adenovirus DNA—the topic that he brought to Cold Spring Harbor in 1979 as a postdoctoral fellow.

4. Tjian R., Fey G., and Graessmann A. 1978. Biological activity of purified simian virus 40 T antigen proteins. *Proc. Natl. Acad. Sci.* **75:** 1279–1283. See also Tjian, page 119, this volume.

5. Stillman B.W. and Gluzman Y. 1985. Replication and supercoiling of simian virus 40 DNA in cell extracts from human cells. *Mol. Cell. Biol.* **5:** 2051–2060.

6. Li J.J. and Kelly T.J. 1984. Simian virus 40 DNA replication in vitro. *Proc. Natl. Acad. Sci.* **81:** 6973–6977.

7. Li J.J. and Kelly T.J. 1985. Simian virus 40 DNA replication in vitro: Specificity of initiation and evidence for bidirectional replication. *Mol. Cell. Biol.* **5:** 1238–1246.

8. Prelich G. and Stillman B.W. 1986. Functional characterization of thermolabile DNA-binding proteins that affect adenovirus DNA replication. *J. Virol.* **57:** 883–892.

9. In the late 1970s, John Smart came to Cold Spring Harbor as a visiting scientist from the Imperial Cancer Research Fund and was soon appointed to lead the Protein Chemistry group at the Laboratory. In 1981, he moved to Biogen's new laboratories in Cambridge, Massachusetts, and has worked in the biotechnology industry ever since.

10. Evans T., Rosenthal E.T., Youngblom J., Distel D., and Hunt T. 1983. Cyclin: A protein specified by maternal mRNA in sea urchin eggs that is destroyed at each cleavage division. *Cell* **33:** 389–396.

11. See footnote 5, Botchan, page 113, this volume.

12. Nakamura R.M. and Tan E.M. 1983. Autoantibodies to nonhistone nuclear antigens and their clinical significance. *Hum. Pathol.* **14:** 392–400.

13. Mathews M.B., Berstein R.M., Franza B.R. Jr., and Garrels J.I. 1984. Identity of the proliferating cell nuclear antigen and cyclin. *Nature* **309:** 374–376.

14. Tan C.K., Castillo C., So A.G., and Downey K.M. 1986. An auxiliary protein for DNA polymerase-delta from fetal calf thymus. *J. Biol. Chem.* **261:** 12310–12316.

NATURE VOL. 326 2 APRIL 1987
———————————ARTICLES———————————
471

The cell-cycle regulated proliferating cell nuclear antigen is required for SV40 DNA replication *in vitro*

Gregory Prelich, Matthew Kostura, Daniel R. Marshak, Michael B. Mathews & Bruce Stillman*

Cold Spring Harbor Laboratory, PO Box 100, Cold Spring Harbor, New York 11724, USA

Cell-free extracts prepared from human 293 cells, supplemented with purified SV40 large-T antigen, support replication of plasmids containing the SV40 origin of DNA replication. A cellular protein ($M_r \sim 36,000$) that is required for efficient SV40 DNA synthesis in vitro has been purified from these extracts. This protein is recognized by human autoantibodies and is identified as the cell-cycle regulated protein known as proliferating cell nuclear antigen (PCNA) or cyclin.

PREREQUISITE to understanding how cellular proliferation is controlled in eukaryotes is the identification of proteins that are directly involved in DNA replication, particularly those involved in its regulation. The transition into the replicative S phase of the cell cycle is accompanied by induction of a specific set of cellular proteins, some of which probably participate directly in DNA replication. A great deal has been learned about the enzymatic requirements for DNA replication in prokaryotes, largely because of the availability of well-defined phage and cellular origins of replication and of cell-free extracts that are capable of specifically replicating DNA molecules containing those origins[1-3]. An analogous approach has been adopted in an effort to elucidate the mechanism of eukaryotic DNA replication. Little is known about the DNA sequences that constitute origins of replication in cellular chromsomes, so viral replicons have become the systems of choice. The development of soluble extracts capable of replicating adenovirus[4], simian virus 40 (SV40)[5-8] and polyomavirus[9] DNAs in vitro has enabled rapid progress to be made in this field. The majority of the factors required for papovavirus DNA replication are of cellular origin and it is this dependence upon host proteins that makes these systems such attractive models to investigate the mechanism of cellular chromosomal replication.

SV40 contains a covalently closed DNA genome of 5,243 base pairs that is maintained in a chromatin structure similar to the structure of cellular chromosomes. Within the minichromosome lies a single origin of replication (*ori*) from which synthesis procedes bi-directionally and semi-conservatively around the molecule[10-12]. The only virus-encoded protein that is required for replication is the SV40 large tumour antigen (large-T antigen)[13], a multifunctional phosphoprotein of relative molecular mass 90,000 (M_r 90K) which is necessary for a number of processes, including transcriptional control[14-16] and cellular transformation[17] as well as DNA replication[18]. Among its biochemical activities are its origin-specific DNA-binding property[19] and an ATP-dependent DNA helicase function[20] that is associated with an ATPase activity. Mutant large-T antigens that are defective for either of these activities are replication defective[20-23]. Cytosolic extracts prepared from monkey or human cells support efficient SV40 DNA replication which is dependent upon purified large-T antigen and covalently closed circular plasmids containing the SV40 *ori* sequence, and closely resembles the mode of DNA synthesis observed in infected cells[5-8]. Addition of a nuclear extract enables the replicating DNA to be concomitantly assembled into minichromosomes[24]. By fractionating the crude cellular cytosolic extract, we have purified a cellular factor that is required for SV40 DNA replica-

tion *in vitro* and have identified it as the proliferating cell nuclear antigen (PCNA)[25-27], a protein recognized by antibodies from some patients with the autoimmune disease, systemic lupus erythematosus (SLE). PCNA, also known as cyclin[28], is a cell-cycle regulated protein that is expressed at elevated levels in transformed and tumour cell lines, and indirect studies have associated it with cellular DNA replication. Thus, this report establishes a direct biochemical link between a cell-cycle regulated, transformation-sensitive protein and the process of DNA replication. An accompanying paper shows that this factor stimulates the activity of the DNA polymerase-δ enzyme[29].

Cellular replication proteins

Our strategy for identification of the cellular replication factors is to divide the cytosolic extract derived from human 293 cells[7,30] into multiple fractions, each of which will ultimately contain a single replication factor that can be purified. The post-ribosomal supernatant (S100) fraction was initially divided by phosphocellulose column chromatography into two components (fractions I and II, Fig. 1a). In standard reactions containing large-T antigen and origin-containing plasmid DNA, each of the these fractions alone was inactive, but efficient replication was restored when they were incubated together (Fig. 1b). Fraction I was further divided by DEAE-cellulose column chromatography into two components, A and B (Fig. 1a). The combination of fractions A and II gave a low level of DNA replication (Fig. 1b), which was dependent upon the presence of large-T antigen and a functional *ori* sequence as expected (data not shown and ref. 29). Addition of fraction B stimulated DNA synthesis greatly, restoring it to a level comparable to that given by the unfractionated cytosol S100 fraction.

We initially concentrated on fraction B because early results indicated that this fraction contains a single, essential replication component. Its purification was followed by its ability to stimulate replication in the presence of optimal amounts of fractions II and A, as measured by the incorporation of ^{32}P-dAMP into acid-precipitable material. Using this assay, the stimulatory activity present in fraction B was isolated by sequential phenyl-Sepharose column chromatography, Mono Q column chromatography and glycerol gradient velocity sedimentation (Table 1). In other experiments, the replication factor failed to bind to native and denatured DNA cellulose columns under a variety of conditions (data not shown). The purified protein obtained from a preparative glycerol gradient was diluted and re-run on an analytical glycerol gradient (Fig. 2). Fractions from the gradient were examined by SDS-polyacrylamide gel electrophoresis (PAGE) followed by staining with Coomassie brilliant blue dye (Fig. 2a), and were assayed for replication activity (Fig. 2b). The peak of replication activity corresponds exactly

---
* To whom correspondence should be addressed.

Bell S.P. and Stillman B. 1992. **ATP-dependent recognition of eukary-otic origins of DNA replication by a multiprotein complex.** *Nature* **357:** 128–134. (Reprinted, with permission, from Macmillian Publishing Ltd ©1992.)

I N 1986, I WAS INVITED TO PRESENT A SEMINAR at Stanford University in the Department of Biochemistry, then arguably the home of DNA replication research. While there, I met Ron Davis who had in 1979 along with others reported that the yeast *Saccharomyces cerevisiae*[1] could be transformed with high efficiency by plasmids that contained DNA sequences called autonomously replicating sequences (ARSs).[2] These are short DNA elements that are capable of directing the replication of neighboring DNA sequences. I was surprised to learn that Davis' laboratory had no plans to continue to work on ARS plasmids, and so on the plane ride back to New York, I began to plan a series of experiments that resulted in my laboratory beginning studies of chromosome replication in yeast.

If, as seemed likely, ARSs were sites at which DNA synthesis was initiated, then yeasts should contain protein(s)—"initiators"—that would bind to ARS elements in a sequence-specific manner. In my laboratory at the time was John Diffley who was working on adenovirus DNA replication. We discussed the possibility of searching for proteins that could bind specifically to the ARS consensus sequence,[3] in the hope that some of these proteins might be targeted to the yeast origins of DNA replication and might be required for initiation of DNA synthesis—in other words, cellular proteins that would be the functional equivalents of SV40 T antigen. John succeeded in isolating a couple of ARS-binding proteins with interesting biochemical properties, but neither turned out to be the key initiator protein. This result was to be repeated several more times as successive students and postdoctoral fellows tried their luck at the problem and then moved on to more tangible projects.

In the meantime, York Marahrens, a graduate student, undertook a project to search for the DNA sequences that promoted ARS activity. The existing literature suggested that the ARS consensus sequence was necessary but not sufficient for origin

---

BRUCE STILLMAN (*See biographical footnote on p. 41.*)

activity, but attempts to define the additional sequences had so far failed. York employed a technique known as linker scanning mutagenesis[4] which, when combined with the sensitive assay that we had designed to test the mutants, resulted in the identification of four specific elements in the *ARS1* origin, one of which was the 11-base-pair ARS core consensus sequence.[5] Interestingly, the other three elements—the B elements—were partially redundant, a novelty in origin sequences at that time.

When Steve Bell joined my laboratory in 1991, he set out to develop a cell-free system that contained all the protein machinery required for the process of DNA replication and was dependent on ARSs to initiate DNA synthesis. Others in my lab had tried and failed, but the potential payoff was so enormous that we both thought the approach was worth another shot, using the knowledge we had gained from the SV40 DNA replication system. Steve included ATP and Mg^{++} in his assays, cofactors that we knew were essential for cell-free replication of SV40 DNA. Having not much luck with seeing ARS-dependent DNA replication, he used a deoxyribonuclease 1 (DNase 1) protection assay to determine if there were any proteins binding to the ARSs, but none were found until a column fraction yielded a DNase 1 hypersensitive site in one of the B elements identified by Marahrens. We were interested in the hypersensitive site since Toma and Simpson[6] had reported the existence of a nuclease hypersensitive site in *ARS1* DNA in vivo that varied during the cell division cycle. Steve immediately put this activity over another column and, dramatically, a clear protection of origin sequences could be observed, including the essential ARS A element.

Within a short period, a six-subunit protein was purified to homogeneity. We called the protein the origin recognition complex (ORC) after some discussion of alternative names. The use of Marahren's ARS elements and newly created point mutations in the ARS consensus sequence convinced us that we had the right protein. At long last, the cellular initiator protein had been identified. We later demonstrated the existence of conserved proteins in other species including humans, and it is now known that ORC mediates the assembly of a multiprotein replication machine at all origins of DNA replication before the initiation of DNA synthesis.

In all eukaryotic cells—from yeast to humans—chromosomal replication is carefully coordinated with the cycle of cell division. A process of such temporal and spatial complexity necessarily involves many levels of regulation, chief among which is the selection of an initiation site on DNA and assembly of the appropriate enzymatic machinery. The discovery of ORC was therefore an important step in understanding how DNA replication is controlled and has stimulated work in many laboratories, pleasingly including the independent laboratories of John Diffley and Steve Bell. Furthermore, we now know that ORC is required for a number of activities in human cells, including maintenance of heterochromatin, and in centrosome and centromere function.

Notes and References

1. Stinchcomb D.T., Struhl K., and Davis R.W. 1979. Isolation and characterisation of a yeast chromosomal replicator. *Nature* **282:** 39–43.

2. An ARS is a sequence of DNA from the genome of the yeast (*S. cerevisiae*) that, when cloned into a plasmid, promotes transformation with high efficiency and the ability to replicate autonomously in synchrony with cellular chromosomes.

3. The consensus ARS sequence was identified by comparing the sequences of a number of ARS elements. Broach J.R., Li Y.Y., Feldman J., Jayaram M., Abraham J., Nasmyth K.A., and Hicks. J.B. 1983. *Cold Spring Harbor Symp. Quant. Biol.* **47:** 1165–1173 (Vol. 2).

4. McKnight S.L. and Kingsbury R. 1982. Transcriptional control signals of a eukaryotic protein-coding gene. *Science* **217:** 316–324.

5. Marahrens Y. and Stillman B. 1992. A yeast chromosomal origin of DNA replication defined by multiple functional elements. *Science* **255:** 817–823.

6. Thoma F. and Simpson R.T. 1985. Local protein-DNA interactions may determine nucleosome positions on yeast plasmids. *Nature* **315:** 250–252.

ATP-dependent recognition of eukaryotic origins of DNA replication by a multiprotein complex

Stephen P. Bell & Bruce Stillman

Cold Spring Harbor Laboratory, Cold Spring Harbor, New York 11724, USA

A multiprotein complex that specifically recognizes cellular origins of DNA replication has been identified and purified from the yeast *Saccharomyces cerevisiae*. We observe a strong correlation between origin function and origin recognition by this activity. Interestingly, specific DNA binding by the origin recognition complex is dependent upon the addition of ATP. We propose that the origin recognition complex acts as the initiator protein for *S. cerevisiae* origins of DNA replication.

THE complex process of eukaryotic chromosomal replication must be carefully regulated throughout the cell cycle. It is likely that much of this regulation occurs at the level of the initiation of DNA synthesis. Studies of *Escherichia coli* chromosomal, bacteriophage, and eukaryotic viral DNA replication have resulted in a paradigm for the initiation of bidirectional DNA replication[1,2]. For each of these organisms, the first step is the sequence-specific recognition of the origin of DNA replication by an initiator protein. Binding of the initiator causes partial untwisting of the double helix adjacent to the recognition site and the subsequent action of a helicase leads to further unwinding of the DNA duplex. This unwound DNA structure serves as a template for the initiation of DNA synthesis. Although protein factors likely to be involved in the unwinding and elongation stages of eukaryotic chromosomal DNA replication have been described[3,4], proteins involved in the initial stage of eukaryotic origin recognition have not been identified.

The availability of short well characterized chromosomal origins of DNA replication derived from the yeast *S. cerevisiae* make it a particularly useful organism to study the earliest steps of eukaryotic DNA replication. Originally identified as chromosomal sequences able to confer high frequency of transformation on plasmid DNA[5], a subset of these autonomous replication sequences (ARS) were subsequently shown to act as true origins of replication in the chromosome[6-9]. Sequence comparison of numerous ARS elements led to the identification of an 11-base-pair sequence that is conserved across all ARS elements and is referred to as the ARS core consensus sequence (ACS)[10,11]. Studies of the sequences required for ARS function have found that although the ACS is necessary, sequences 3′ to the T-rich strand of the ACS also are required for ARS function[11]. A detailed analysis of *ARS1* identified four *cis*-acting elements that constitute this chromosomal origin of DNA replication[12]. In addition to an essential A element containing the ACS, three distinct elements within the 3′ region, B1, B2 and B3, are required for efficient *ARS1* function. One of these elements, B3, is a

NATURE · VOL 357 · 14 MAY 1992

RICHARD J. ROBERTS

Klimašauskas S., Kumar S., Roberts R.J., and Cheng X. 1994. **HhaI methyltransferase flips its target base out of the DNA helix.** *Cell* 76: 357–369. (Reprinted, with permission, from Elsevier ©1994.)

FOR THOSE OF US WHO STUDY RESTRICTION ENZYMES and try to clone their genes, their *alter egos*—the DNA methyltransferases that necessarily accompany them—have proved equally interesting. Thus, while cloning and sequencing the genes coding for restriction enzymes, the sequences of the cognate DNA methyltransferases were also determined. My group at Cold Spring Harbor had spent a great deal of time developing software for analyzing and assembling DNA sequences, and we wanted to use these techniques to compare the sequences of restriction enzymes and DNA methyltransferases. It soon became clear that restriction enzyme sequences were all very different from one another and were also quite different from the sequences of any other proteins that were known. In contrast, the DNA methyltransferases showed a lot of sequence similarity, reflecting the fact that they all use *S*-adenosylmethionine as their substrate and catalyze common chemical reactions to attach the methyl group to a specific base in DNA.

In the case of the methyltransferases that form 5-methylcytosine in DNA, the attachment process is not simple. Dan Santi and his colleagues postulated and then provided biochemical evidence for an initial step in the reaction involving the formation of a covalent bond between the methyltransferase and the cytosine base in DNA that was to be methylated.[1] From a structural standpoint, this poses something of a challenge. In double-stranded DNA, the bases lie inside the helix, permitting them to base pair, whereas the outside of the helix is formed by the phosphodiester

RICH ROBERTS is currently the Chief Scientific Officer at New England Biolabs, Ipswich, Massachusetts. He was educated in England, attending the University of Sheffield where he obtained a B.Sc. in Chemistry in 1965 and a Ph.D. in Organic Chemistry in 1968. His postdoctoral research was carried out in Professor J.L. Strominger's laboratory at Harvard. At Cold Spring Harbor, he began work on the newly discovered Type II restriction enzymes in 1972, and in the next few years, more than 100 such enzymes were discovered and characterized in his laboratory. Rich was awarded the Nobel Prize in Physiology or Medicine in 1993 for his part in the discovery of mRNA splicing in 1977. roberts@neb.com

Rich Roberts, 1992. *(Photograph by Margot Bennett, courtesy of Cold Spring Harbor Laboratory Archives.)*

backbone. For a methyltransferase first to recognize a specific DNA sequence and then to form a covalent bond and perform chemistry on an individual base would seem to require a considerable distortion in the DNA.

We knew that some transcription factors and other DNA-binding proteins could bend and kink the DNA, and it became of interest to find out exactly what sort of distortion the methyltransferases might induce. With this in mind, I recruited a postdoctoral fellow, Ashok Dubey, fresh from his Ph.D. in India, and he set about working out a purification scheme for the DNA methyltransferase, M.MspI, with a view to obtaining crystals of the enzyme. The actual crystallization experiments were carried out in collaboration with Xiaodong Cheng, another young postdoctoral fellow in Jim Pflugrath's laboratory at Cold Spring Harbor.[2] Ashok was able to construct a strain of bacteria that overexpressed M.MspI and could make highly purified material. However, attempts to crystallize the methyltransferase, either alone or in combination with *S*-adenosylmethionine or DNA, proved fruitless. No crystals could be found.

After a year of failures, we decided to switch to another methyltransferase, and Ashok dutifully set out to make an overexpressing clone of M.HpaII and to purify quantities sufficient for crystallization. Again, the protein refused to cooperate and no crystals were formed. After three frustrating months, Ashok received a very good offer to return to a job in India, and I guess, fearing that I was about to ask him to try yet another DNA methyltransferase, he took the job. Soon afterward, another postdoctoral fellow, Sanjay Kumar, joined my laboratory and I was able to persuade him to start work on M.HhaI. Within a short time, he had made an over-expressing strain and purified the protein, and this time, again in collaboration with Xiaodong Cheng, crystals were obtained. The structure of the native methyl-transferase bound to its cofactor, S-adenosylmethionine, was published in 1993.[3] But in the absence of DNA, this told us rather little. A visiting scientist in my laboratory, Saulius Klimašauskas,[4] began the job of trying to crystallize a complex of M.HhaI binding to DNA. This also proved possible, and in 1993, a structure was produced in collaboration with Xiaodong Cheng. Amazingly, rather than the crude distortion of DNA represented by a kink or a bend, a very clean distortion had taken place in that the target base had flipped 180° out of the helix and into the pocket of the enzyme where the chemistry could take place. Not only was this an exciting finding, but the timing proved most fortuitous, because I was able to use this discovery as the basis for my Nobel Prize lecture in Stockholm. Since that time, numerous other enzymes that perform chemistry on DNA have also been shown to flip bases out of DNA.

This discovery of base flipping illustrates well both the serendipitous nature of scientific research and the elegance used by Nature to solve problems.

Notes and References

1. Santi D.V., Garrett C.E., and Barr P.J. 1983. On the mechanism of inhibition of DNA-cytosine methyltransferases by cytosine analogs. *Cell* **33:** 9–10.

2. In the early 1990s, Jim Pflugrath held the W.M. Keck Chair in Structural Biology at Cold Spring Harbor. He left the Laboratory in 1994 for Rigaku Americas Corporation, a company specializing in instrumentation for X-ray spectrometry and diffraction, as well as small molecule and protein crystallography. Jim is a member of the Program in Structural Computational Biology and Molecular Biophysics at Baylor College of Medicine, Houston.

3. Cheng X., Kumar S., Posfai J., Pflugrath J.W., and Roberts R.J. 1993. Crystal structure of the HhaI DNA methyltransferase complexed with S-adenosyl-L-methionine. *Cell* **74:** 299–307.

4. Saulius Klimašauskas is currently Head of the Laboratory of Biological DNA Modification at the Institute of Biotechnology, Vilnius, Lithuania.

Cell, Vol. 76, 357–369, January 28, 1994, Copyright © 1994 by Cell Press

HhaI Methyltransferase Flips Its Target Base Out of the DNA Helix

Saulius Klimasauskas,*‡ Sanjay Kumar,†
Richard J. Roberts,† and Xiaodong Cheng*
*W. M. Keck Structural Biology Laboratory
Cold Spring Harbor Laboratory
Cold Spring Harbor, New York 11724
†New England Biolabs
32 Tozer Road
Beverly, Massachusetts 01915

Summary

The crystal structure has been determined at 2.8 Å resolution for a chemically-trapped covalent reaction intermediate between the HhaI DNA cytosine-5-methyltransferase, S-adenosyl-L-homocysteine, and a duplex 13-mer DNA oligonucleotide containing methylated 5-fluorocytosine at its target. The DNA is located in a cleft between the two domains of the protein and has the characteristic conformation of B-form DNA, except for a disrupted G–C base pair that contains the target cytosine. The cytosine residue has swung completely out of the DNA helix and is positioned in the active site, which itself has undergone a large conformational change. The DNA is contacted from both the major and the minor grooves, but almost all base-specific interactions between the enzyme and the recognition bases occur in the major groove, through two glycine-rich loops from the small domain. The structure suggests how the active nucleophile reaches its target, directly supports the proposed mechanism for cytosine-5 DNA methylation, and illustrates a novel mode of sequence-specific DNA recognition.

Introduction

DNA methylation is found in diverse organisms ranging from bacteria to mammals and plants. Cytosine-5-methyltransferases (m5C-MTases) are involved in a variety of biological processes in prokaryotes and eukaryotes by catalyzing the transfer of a methyl group from S-adenosyl-L-methionine (AdoMet) to the C5 position of cytosine. MTases exist as a component of restriction and modification systems in bacteria (see review by Roberts and Halford, 1993). Although the exact function of DNA methylation in eukaryotic cells is not fully understood, DNA methylation has been implicated in the control of a number of cellular processes in eukaryotes, including transcription (Busslinger et al., 1983; Jones, 1985; Cedar, 1988; Boyes and Bird, 1991), genomic imprinting (Reik et al., 1987; Swain et al., 1987), developmental regulation (Antequera et al., 1989), mutagenesis (Cooper and Youssoufian, 1988), transposition (Fedoroff, 1989), DNA repair (Hare

and Taylor, 1985; Brown and Jiricny, 1987, 1988), X inactivation (Gartler and Riggs, 1983; Pfeifer et al., 1990), and chromatin organization (Bird, 1986; Selig et al., 1988; Lewis and Bird, 1991). Aberrations in cytosine-5-methylation may play a role in human genetic disease (Oberle et al., 1991). Recently, the mouse DNA MTase was shown to be essential for normal embryonic development (Li et al., 1992) and to be associated with DNA replication foci (Leonhardt et al., 1992).

All m5C-MTases share a set of well-conserved primary sequence motifs (Posfai et al., 1989; Lauster et al., 1989) that are believed to reflect the common functions required for DNA methylation. As originally proposed by Wu and Santi (1985), the catalytic mechanism of m5C-MTases involves nucleophilic attack on C6 of the target cytosine by a conserved cysteine residue to generate a covalent intermediate (Figure 1). This results in the addition of a methyl group to the C5 position of cytosine followed by elimination of the C5 proton and release of the covalent intermediate. This intermediate can be trapped if the target cytosine base has a fluorine atom substituted for a hydrogen atom at its C5 position (Osterman et al., 1988). Direct support for this mechanism has been provided by the isolation and characterization of covalent intermediates formed by M. HaeIII (Chen et al., 1991), M. EcoRII (Friedman and Ansari, 1992), and the human methyltransferase (Smith et al., 1992). Replacement of the corresponding cysteine with other amino acids in several m5C-MTases abolishes catalytic activity (Wilke et al., 1988; Wyszynski et al., 1992; Mi and Roberts, 1993; Chen et al., 1993).

M. HhaI is an m5C-MTase that forms part of a type II restriction–modification system from *Haemophilus haemolyticus*. The enzyme recognizes the specific tetranucleotide sequence, 5′-GCGC-3′ (Roberts et al., 1976) and methylates the first cytosine residue (Mann and Smith, 1979). The crystal structure of a binary complex of M. HhaI with AdoMet has revealed the relationship between the tertiary structure and the six most highly conserved sequence motifs (Cheng et al., 1993). The structure does not have any recognizable DNA-binding motifs (Harrison, 1991; Phillips, 1991; Pabo and Sauer, 1992). Rather, it folds into two domains separated by a large cleft, which was predicted to be the site for DNA binding (Cheng et al., 1993). Most of the invariant residues are located within the loop regions facing the cleft from the large domain, which provides the cofactor-binding site and essential catalytic residues. The small domain contains the variable region that is responsible for the specific recognition of target DNA (Balganesh et al., 1987; Klimasauskas et al., 1991) and the selection of the base to be methylated (Mi and Roberts, 1992).

We now report the crystal structure of a covalent ternary complex between M. HhaI, a 13-mer oligonucleotide containing its recognition sequence with 5-fluorocytosine at the site of methylation, and the reaction product S-adenosyl-L-homocysteine (AdoHcy).

‡Permanent Address: Institute of Biotechnology FERMENTAS, 232028 Vilnius, Lithuania.

CELL BIOLOGY

From Architecture to Functional Analysis

DAVID L. SPECTOR

A LTHOUGH IT IS OFTEN SAID THAT "A PICTURE IS WORTH a thousand words," this rings no truer than in the field of cell biology where the microscope has provided many amazing images of cellular organization and dynamics. It is no surprise that Jim Watson would fully appreciate this area of science as he has always been one who derives both insight and pleasure from images. Jim's long-time interest in fine art, the remarkable image of the X-ray diffraction pattern of DNA, his passionate attention to the detail of where each tree should be planted or sculpture placed on Laboratory grounds, and his engagement with how each building is designed all attest to his fascination with "the image" and how it communicates a complex story in a concise and logical manner.

The area of cell biology at Cold Spring Harbor Laboratory can, however, be traced back before Jim's time—back to 1958 when Berwind Kaufmann acquired an RCA EMU 3C transmission electron microscope to study the fine structure of cellular components. Subsequently, in the 1970s, the cell biology efforts at the Lab were associated with the "Mammalian Cell Genetics" and "Proteins in SV40-infected and -transformed Cells" programs. During this time, studies by Robert Pollack and his associates focused on growth con-

---

DAVID L. SPECTOR, who holds a Ph.D. from Rutgers University, joined Cold Spring Harbor Laboratory in 1985. Previously an Assistant Professor at Baylor College of Medicine, Houston, he is currently Director of Research at Cold Spring Harbor and a Professor in the Watson School of Biological Sciences. He is an editor of several laboratory manuals including *Basic Methods in Light Microscopy* and *Live Cell Imaging*.

trol and understanding the differences between normal and transformed cells. Some would say that Jim's interest in cell biology was initiated at the 1972 CSHL Symposium on "The Mechanism of Muscle Contraction" which featured a session on "Contractile Proteins in Nonmuscle Tissue." It was after that symposium that Jim invited Robert Goldman, from Carnegie Mellon Institute, to spend a year at CSHL to boost the microscopy effort. During this time (1973), Robert Goldman and Robert Pollack acquired funding to purchase a scanning electron microscope (SEM) allowing them to study cell surface phenomena related to cellular transformation. However, once Jim made CSHL his full-time focus in 1974, cell biology was expanded. The group was soon joined by Guenter Albrecht-Buehler, a physicist turned biologist, who was instrumental in getting the newly acquired SEM up and running, and who later focused his efforts on cell motility and symmetry (see below).

As described in the first paper in this section, Elias Lazarides and Klaus Weber[1] developed an anti-actin antibody using sodium-dodecyl-sulfate-denatured protein, circumventing the poor immunogenicity of the native protein (see also Weber, p. 61). With advice from Robert Goldman and Robert Pollack, they went on to demonstrate, using immunofluorescence in nonmuscle cells, that actin exists as dramatically patterned filamentous structures. This study provided a global picture of actin distribution throughout the cytoplasm, and although immunofluorescence was not a new technique, the excitement generated by these images in essence represented a rebirth of immunofluorescence for cell biologists. The images from this study also made a significant impact on Jim, and he was looking for the latest picture each night when he made his habitual "rounds" through the labs. Furthermore, when transformed cells were examined using the same immunofluorescence assay, they appeared to have a reduced number of bundles of actin filaments. Thus, these data fitted beautifully into the ongoing theme of Robert Pollack's "Mammalian Cell Genetics" program: to identify differences between normal and transformed cells. This excitement was carried on to the 1975 CSHL meeting on "Cell Motility" which focused on the structural and molecular aspects of motility in nonmuscle cells with emphasis on the microfilament and microtubule systems. A series of papers from this meeting resulted in the classic publication *"Cell Motility"*—a three-volume set of "green books" edited by Robert Goldman, Tom Pollard, and Joel Rosenbaum.[2] Many would say that this meeting was a critical event in the initiation of the cytoskeleton field. At about this time, Keith Burridge and Lan Bo Chen joined the laboratory and carried out insightful studies producing many beautiful images of the localization of α-actinin, fibronectin, and actin. The finding by Lan Bo that the distribution of fibronectin outside the cell paralleled the pattern of actin filaments in the cytoplasm in normal cells, but not in cancer cells, was a highlight of many discussions.

Studies on the cytoskeleton were intimately linked to those on cell migration. The second paper in this section describes a technique developed by Guenter Albrecht-Buehler to visualize and track the movements of mammalian cells attached to gold-particle-coated coverslips. In this 1977 paper, Guenter[3] coined the term "phagokinetics" to describe the movements that he visualized by darkfield microscopy (see also Albrecht-Buehler, p. 65). He found differences in the track patterns among different cell types; he also found that "daughter cells performed directional changes in mirror symmetrical or identical ways provided the cells did not encounter other cells." These observations led him to posit that the behavior of a cell is predetermined at mitosis. Numerous other studies around the same time identified various levels of symmetry between postmitotic daughter cells. However, the principles underlying these observed symmetries are still unclear.

Shortly thereafter, Stephen Blose, who joined the laboratory in 1978, discovered a ring of intermediate filaments surrounding the nuclei of vascular endothelial cells, and Jim Garrels[4] joined the program and began studying the differences in protein composition between normal and transformed cells using large two-dimensional gels and computer analysis. As part of this effort, the *QU*antitative *E*lectrophoresis *ST*andardized in 2 *D*imensions (QUEST-2D) laboratory was initiated (see also Garrels, *Techniques*, p. 219).

In 1979, the Cell Biology group, previously scattered throughout the laboratory, united in their newly renovated home in the McClintock Building. Jim Feramisco, arriving in 1978, focused his efforts on α-actinin and together with Keith Burridge made a remarkable observation suggesting that vinculin (then called "focin," a name suggested by Jim Watson), because of its localization to the tips of actin microfilament bundles and underlying the matrix of fibronectin, might link the actin cytoskeleton and the extracellular matrix.

An abundance of cytoskeletal proteins had by this time been localized in fixed cells by immunofluorescence. But what of living cells? Being able to localize specific proteins in living cells would not only allow one to determine if the fixed cell studies were correct, but also enable one to assess the dynamics of the respective proteins. Jim Feramisco's laboratory was among those that popularized microinjection of fluorescently conjugated proteins into living mammalian cells, a technique he learned from Adolf Graessmann who was visiting the Lab from the Freie University in Berlin. Using this approach, Steve Blose and Jim were able to show that the injected cytoskeletal proteins maintained their native localization after fixation and permeabilization. Together with Jim Lin, they also microinjected monoclonal antibodies to assess potential cellular defects resulting from the formation of specific antigen/antibody complexes. Thus, the combination of biochemistry, immunology, and microscopy was beginning to be used in a coordinated fashion to assess function in living cells. The theme of cell biology and the cytoskeleton continued with the 1981 CSHL Symposium on Quantitative Biology, which focused on "The Organization of the Cytoplasm." In his meeting Summary, Bill Brinkley referred to the meeting as "the first major symposium in the 20th century to embrace the topic of cytoplasm as a holistic entity."[5]

The evolution of cell biology at CSHL closely paralleled the research on cancer. Thus, it was no surprise that the function of the first identified oncogene (*ras*) was quickly examined by the cell biologists. As reported in the third paper in this section, published in 1985, Dafna Bar-Sagi and Jim Feramisco[6] showed that microinjection of the human H-*ras* oncogenic protein, but not the proto-oncogene, promoted the unexpected differentiation of PC12 cells into neuron-like cells in the absence of nerve growth factor (see also Bar-Sagi and Feramisco, p. 69). This was in striking contrast to earlier studies showing that Ras microinjected into NIH-3T3 and NRK cells promoted proliferation in the absence of growth factors. Clearly, *ras* could promote proliferation or differentiation, depending on cellular context.

During this period, significant emphasis was also placed on understanding the heat shock (or stress) response through the studies of Bill Welch. David Helfman, together with Stephen Hughes and Jim Feramisco, developed a method for the immunological screening of cDNA expression libraries, a method they used to characterize the nonmuscle tropomyosin genes. Fumio Matsumura studied the cell biology of these tropomyosins and other actin-binding proteins. Also at this time, Robert Franza and Jim Garrels initiated efforts to examine the effect of oncogene expression on cellular proteins using the high-resolution two-dimensional gel approach developed in the QUEST-2D lab.

While studies of cytoplasmic structure/function continued, cell biology at CSHL expanded its scope to include the nucleus with my arrival in 1985 and the arrival of Carol Greider in 1988. As a CSHL Fellow, Carol continued her work on telomeres and telomerase, work she had initiated while a graduate student in Liz Blackburn's lab at Berkeley. The fourth paper in this section highlights Carol's efforts to examine the relationship between telomere length and aging[7] (see also Greider, p. 75). Using Southern blots to examine telomere length in human fibroblasts grown in culture for different numbers of cell doublings, she showed that the amount and length of telomeric DNA decrease as a function of serial passage, i.e., she showed that older cells had shorter telomeres. This study provided a potential mechanism by which primary cells could control their number of doublings ("Hayflick number") in cell culture and had a significant impact on many fields.

Upon my arrival in 1985, and with Jim's interest in microscopy reinvigorated, a new microscopy facility was established at the Lab. The existing microscopes were functional but relatively old and the field was on the verge of a major revolution: The first commercial confocal microscope was about to be released. Acquisition of a new transmission electron microscope and image analysis system enabled us to examine the localization of small nuclear ribonucleoprotein particles (snRNPs), essential components of the pre-mRNA splicing machinery, in nuclear speckles (also known as interchromatin granule clusters). As described in the fifth paper in this section, we took the laborious approach of combining immunoelectron microscopy with serial section reconstruction[8] (see also Spector, p. 83). Our study and model of the cell nucleus initiated fierce debate and put "nuclear speckles" on the map. The suggestion that "a new set of organizational criteria that differentiates functional compartments within the nucleus may be needed" has clearly stood the test of time. The initiation of a biennial CSHL meeting on "Dynamic Organization of Nuclear Function" in 1998 represented the first major international meeting to bring together investigators working across the entire field of nuclear structure and function and coincided with major technical innovations.

Today, cell biology continues to thrive at CSHL where cytoplasmic and nuclear constituents are now routinely followed over time in three dimensions in living cells, and the ability to access molecular composition and the function of cellular constituents keeps the pulse of cell biology moving at a rapid pace. If an "image" is worth a thousand words, a "movie" is likely worth a million.

Acknowledgments

I greatly appreciate and enjoyed discussions with Steve Blose, Bill Brinkley, Guenter Albrecht-Buehler, Keith Burridge, Robert Goldman, and Jim Feramisco. I have tried to place the papers, selected by the editors for this section, into the overall context of Cell Biology at CSHL during the described time period. This Introduction is not meant to be a comprehensive review and I apologize to those whose studies I was not able to highlight.

Notes and References

1. Lazarides E. and Weber K. 1974. Actin antibody: The specific visualization of actin filaments in non-muscle cells. *Proc. Natl. Acad. Sci.* **71:** 2268–2272.

2. Goldman R.D., Pollard T.D., and Rosenbaum J.L., eds. 1976. *Cell motility.* Cold Spring Harbor Laboratory Press, Cold Spring Harbor, New York. 1404 pp.

3. Albrecht-Buehler G. 1977. Daughter 3T3 cells. Are they mirror images of each other? *J. Cell Biol.* **72:** 595–603.

4. Garrels J.I. 1989. The QUEST System for quantitative-analysis of two-dimensional gels. *J. Biol. Chem.* **264:** 5269–5282.

5. Brinkley B.R. 1982. Summary: Organization of the cytoplasm. *Cold Spring Harbor Symp. Quant. Biol.* **46:** 1029–1040.

6. Bar-Sagi D. and Feramisco J.R. 1985. Microinjection of the ras oncogene protein into PC12 cells induces morphological differentiation. *Cell* **42:** 841–848.

7. Harley C.B., Futcher A.B., and Greider C.W. 1990. Telomeres shorten during ageing of human fibroblasts. *Nature* **345:** 458–460.

8. Spector D. 1990. Higher order nuclear organization: Three-dimensional distribution of small nuclear ribonucleoprotein particles. *Proc. Natl. Acad. Sci.* **87:** 147–151.

Lazarides E. and Weber K. 1974. **Actin antibody: The specific visualization of actin filaments in non-muscle cells.** *Proc. Natl. Acad. Sci.* **71:** 2268–2272. (Reprinted, with permission.)

A CTIN HAS BEEN AN OBJECT OF SCIENTISTS' curiosity for 120 years. The first credible experimental descriptions of the protein's properties were made in 1887[1]—two years before the first biological laboratories at Cold Spring Harbor were established.[2] But eight decades were to pass before the Laboratory took an active interest in the molecule. For most of the intervening time, actin was believed to be primarily a protein of muscle cells where, complexed with myosin and provided with a source of energy, it functioned in muscle contraction. However, by the late 1960s, chiefly as a result of electron microscopic work in Howard Holtzer's group at the University of Pennsylvania,[3] actin filaments had been detected not just in muscle cells, but in a wide variety of mammalian cell types. There things stalled for several years, chiefly because of technical limitations: High-resolution electron microscopy was both slow and restrictive. Building a picture of the distribution of actin throughout a single cell would require taking and assembling many high-resolution electron micrographs. What was needed to break the impasse was a sensitive and specific reagent that would provide an overview of the distribution of actin in many cells simultaneously while preserving the protein's intracellular distribution.

As Elias Lazarides has pointed out,[4] "the obvious choice would have been localization through antibodies; except that, in the past, a number of investigators had tried to raise antibodies against actin purified by conventional techniques but had failed, resulting only in the generation of antibodies to other antigens contaminating the actin preparation. This was presumably because actin was highly conserved and hence poorly immunogenic in its native state."

KLAUS WEBER, originally from Poland, earned his undergraduate (1962) and graduate (1964) degrees from the University of Freiburg. He came to the United States to work with Jim Watson at Harvard University and eventually ran a joint laboratory with Jim and Wally Gilbert. He came to Cold Spring Harbor in 1972 on leave from Harvard, where he was by then a Professor of Biochemistry. In 1975, he moved to the Max-Planck-Institute in Göttigen, to head a new section on cell biochemistry. office.weber@mpibpc.gwdg.de

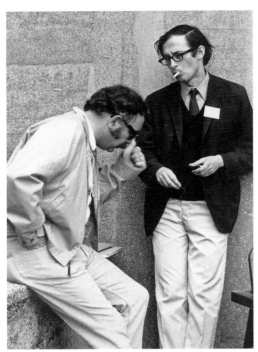

Gunter von Ehrenstein and Klaus Weber, 1970.
(Courtesy of Cold Spring Harbor Laboratory Archives.)

To solve the problem, we decided to use denaturing polyacrylamide gels as the final step of purification of actin. As the accompanying paper describes, actin from SV40-transformed mouse fibroblasts was first enriched by standard biochemical fractionation and then further purified by SDS gel electrophoresis. Simple and quick, we hoped that this method would provide a general solution to the problems of purifying cellular components as antigens, hence facilitating the generation of antibody probes of high specificity and sensitivity. However, in 1974, a major limitation was the resolving power of the gels, which meant that the method could be effectively used only to purify proteins that, like actin, were abundant components of cells. But within a short period of time, super high-resolution two-dimensional gels became available that extended the range of the technique to proteins of much lower abundance. Clearly, a cell biologist no longer had to become expert in protein chemistry before isolating a protein to be used as antigen. Later technical advances involved the use of monoclonal antibodies, which eliminated any lingering doubts about the specificity of the immunological probes; antibodies raised in different species of animals allowed cells to be probed simultaneously with antibodies raised against different proteins; and finally, improvements in fluorescence microscopy, such as confocal imaging, allowed the location of different proteins to be mapped with accuracy.

Our first paper, which demonstrated that actin in cells is organized into a two-dimensional matrix of filamentous structures, was quickly followed by others showing that the filamentous bundles coincided with the classically known microfilament bundles[5] and contained a myriad of other known muscle proteins such as myosin,[6] tropomyosin,[7] and α-actinin.[8]

Today, all this sounds rather mundane. But at the time, no one would have predicted that the cytoskeleton was composed of such a rich array of proteins. Nor could they have anticipated the sheer beauty of the fluorescent images: the tensile tracery of filamentous bundles spanning the length of the cell; other filaments converging on focal points. These images combine two kinds of beauty: the esthetic and the scientific, with the unknown now made visible and the visible now made understood. Stephen Dedalus, the semi-autobiographical character in James Joyce's *Portrait of the Artist as a Young Man*, put it well:

> *Plato, I believe, said that beauty is the splendour of truth. I don't think that it has a meaning but the true and the beautiful are akin.*

> *...although the same object may not seem beautiful to all people, all people who admire a beautiful object find in it certain relations which satisfy and coincide with the stages themselves of all esthetic apprehension. These relations of the sensible, visible to you through one form and to me through another, must be therefore the necessary qualities of beauty.*

Notes and References

1. Halliburton W.D. 1887. On muscle plasma. *J. Physiol.* **8:** 133.

   Finck H. 1968. On the discovery of actin. *Science* **160:** 332.

2. In 1889, John D. Jones gave land and buildings—formerly part of the Cold Spring Whaling Company on the southwestern shore of Cold Spring Harbor—to the newly established Biological Laboratory. The first course at the new Laboratory, a general course on Biology, was taught the following summer.

3. Ishikawa H., Bischoff R., and Holtzer H. 1969. Formation of arrowhead complexes with heavy meromyosin in a variety of cell types *J. Cell Biol.* **43:** 312–328.

4. Lazarides E. 1983. Current contents: This week's citation classic. *Life Sci.* **29:** 19.

5. Goldman R.D., Lazarides E., Pollack R., and Weber K. 1975. The distribution of actin in non-muscle cells. The use of actin antibody in the localization of actin within the microfilament bundles of mouse 3T3 cells. *Exp. Cell Res.* **90:** 333–344.

6. Weber K. and Groeschel-Stewart U. 1974. Myosin antibody: The specific visualization of myosin containing filaments in non-muscle cells. *Proc. Natl. Acad. Sci.* **71:** 4561–4565.

7. Lazarides E. 1975. Tropomyosin antibody: The specific localization of tropomyosin in non-muscle cells. *J. Cell Biol.* **65:** 549–561.

8. Lazarides E. and Burridge U. 1975. α-Actinin: Immunofluorescent localization of a muscle structural protein in non-muscle cells. *Cell* **6:** 289–298.

Reprinted from

Proc. Nat. Acad. Sci. USA
Vol. 71, No. 6, pp. 2268–2272, June 1974

Actin Antibody: The Specific Visualization of Actin Filaments in Non-Muscle Cells

(immunofluorescence/microfilaments/sodium dodecyl sulfate gel electrophoresis)

ELIAS LAZARIDES AND KLAUS WEBER

Cold Spring Harbor Laboratory, Cold Spring Harbor, New York 11724

Communicated by J. D. Watson, March 11, 1974

ABSTRACT   **Actin purified from mouse fibroblasts by sodium dodecyl sulfate gel electrophoresis was used as antigen to obtain an antibody in rabbits. The elicited antibody was shown to be specific for actin as judged by immunodiffusion and complement fixation against partially purified mouse fibroblast actin and highly purified chicken muscle actin. The antibody was used in indirect immunofluorescence to demonstrate by fluorescence light microscopy the distribution and pattern of actin-containing filaments in a variety of cell types. Actin filaments were shown to span the cell length or to concentrate in "focal points" in patterns characteristic for each individual cell.**

Eucaryotic cells contain three basic fibrous structures: filaments, microfilaments, and microtubules. These three structures are thought to be intimately involved in the maintenance of cell shape, in cell movement, and in other important cellular functions (1). Microfilaments are thought to contain actin. This assumption is based on the observation that these structures can be selectively decorated with heavy meromyosin, a specific proteolytic fragment of muscle myosin known to interact with muscle actin (2). Furthermore, actin is now now known to exist as a major protein component of a variety of non-muscle cellular types and in each case it has properties markedly similar to those of its muscle counterpart (3–7)*. The major protein of the microtubular system, tubulin, has been isolated and well characterized (9). The basic protein subunit of the filament structure, however, has not so far been identified. Presumptive muscle proteins like myosin (10–13) and tropomyosin (14) have been found in some non-muscle cells. However, their exact distribution within the cell, as well as their specific localization in one of these fibrous structures, is as yet undetermined.

We have developed a relatively simple way of selectively visualizing filamentous structures in the cell by using antibodies made against different structural proteins. The problem of purifying each antigen separately was circumvented by using sodium dodecyl sulfate (SDS) gel electrophoresis. The denatured proteins are antigenic, and the antibody obtained cross reacts with the native protein. Once specificity has been demonstrated, the antiserum obtained can be used in indirect immunofluorescence to visualize the structures in the cell with which the protein is associated.

In this paper we have used mouse fibroblast actin to test this approach. The actin, purified by SDS gel electrophoresis,

was used to obtain an antibody in rabbits. The antiserum obtained was shown by immunodiffusion and complement fixation to be specific for actin. This antibody was then used in indirect immunofluorescence to show that microfilaments are polymers of actin. This technique also enabled us to demonstrate the complex network of actin filaments in a variety of cell types.

MATERIALS AND METHODS

Growth of Cells. Actin was isolated from the cell line SV101, a clone of mouse fibroblast 3T3 cells transformed by Simian virus 40. This transformed cell line was chosen because it grows to a higher saturation density than the parent 3T3 cell line (15). The cells were grown in roller bottles (Vitro Corp.) in Dulbecco's modified Eagle's medium containing 10% calf serum and 50 μg/ml of gentamycin. At confluency, the medium was removed and the cells were washed with phosphate-buffered saline (PBS). The cells were then scraped off the bottles, collected by low speed centrifugation, and stored at $-70°$.

Preparation of Actin. The cells were thawed and homogenized in 20 volumes of 95% ethanol. The precipitate was collected by low speed centrifugation, washed immediately with ether, and air dried. The yield from 10 bottles was approximately 1.2 g of ethanol-ether powder. The ethanol-ether powder was stirred at 4° in 0.01 M sodium phosphate buffer (pH 6.8), 10 mM $MgCl_2$, and 1 mM dithiothreitol (15 ml of buffer per g of powder) for 3–5 hr. The supernatant was made 30% in ammonium sulfate by adding 0.17 g of ammonium sulfate per ml of extract. After stirring for 30 min at 4°, the precipitate was collected by centrifugation and dissolved in and dialyzed against 0.01 M Tris·HCl (pH 7.5), 10^{-4} M $CaCl_2$, 1 mM dithiothreitol, and 10^{-4} M ATP. The actin could be further purified by a second precipitation at 30% ammonium sulfate saturation. Under these conditions, 1 g of ethanol-ether powder yields approximately 1 mg of actin.

Highly purified chicken muscle actin was a generous gift from Dr. Susan Lowey.

Antibody Preparation. The actin used as an antigen in rabbits was purified through SDS slab gel electrophoresis from the high speed supernatant (100,000 $\times$ g) of a mouse fibroblast cell homogenate (see *Results*). Approximately 400 μg of the antigen was injected in complete Freund's adjuvant and 2 weeks later the rabbits were boosted with an additional 400 μg. Blood was collected 6 weeks after the last injection and the serum was clarified by centrifugation at 10,000 rpm. The gamma globulin fraction was partially purified using precipitation with half saturated ammonium sulfate. It was

---

Abbreviations: SDS, sodium dodecyl sulfate; PBS, phosphate-buffered saline.
* The authors apologize for not referring to all the contributors in this field. The reader is referred to a recent detailed review of actin and myosin in non-muscle cells for complete references (8).

GUENTER ALBRECHT-BUEHLER

Albrecht-Buehler G. 1977. **Daughter 3T3 cells. Are they mirror images of each other?** *J. Cell Biol.* **72:** 595–603. (Reprinted, with permission, from Elsevier ©1977.)

———

S YMMETRIES ARE EVERY PHYSICIST'S DELIGHT, and I am no exception. After all, symmetry means that something is invariant against mirror imaging, which, in turn, may lead to a more general invariance, which, in turn, may lead to the discovery of a new law of nature, which, in turn,...I was dreaming along these lines while rocking with one foot the cradle of my recently born son Conrad and staring in disbelief at a bunch of photographs that contained mirror symmetrical tracks left by 3T3 mouse fibroblasts on a gold-dusted glass surface.

Shortly after coming to Cold Spring Harbor, I had found this particular technique of a biological "cloud chamber" in which cells migrate on a glass surface coated with tiny colloidal gold particles. As the cells remove the particles along their way, they, like high-energy particles in a cloud chamber, leave tracks of their migrations. When viewed in a dark-field light microscope, the cells appear as blinding white bodies at the end of velvety black tracks in a dense lawn of warmly golden particles that cover the entire glass surface. A very beautiful sight that no photograph can adequately reproduce.

The technique freed the student of cell motility from focusing, exposing, developing, and measuring miles of 16-mm time-lapse films that, despite all the effort and expense, allowed the recording of the movements of only a handful of cells. Instead, using my tracking technique, one could leave the dish with the cells on gold-coated coverslips in the culture incubator for 1–2 days, fix them, and subsequently study at leisure the tracks of thousands of cells. Besides, the technique was

GUENTER ALBRECHT-BUEHLER, trained as a physicist in Germany in the early seventies, became interested in the motility of cell membranes during his postdoctoral years at the Friedrich Miescher Institute in Basel. He moved to Cold Spring Harbor in 1974 as Head of the Cell Biology group. When he left the Laboratory in 1982, Jim Watson wrote "...Guenter has generated many innovative ideas about cell motility and we shall miss the deep intelligence that he brings to bear on behaviour at the cellular level." Guenter is now Professor of Cell Biology and Anatomy at Northwestern Univerity Medical School, Chicago. g-buehler@northwestern.edu

infinitely cheaper than time-lapse filming, which was an important condition, because I had just recently arrived in the United States and was in no position to ask for much grant support. Of course, the time-lapse filming technique with its small samples of cells that tended to walk out of the frame could never have discovered the symmetries of cell migration. The tracking technique had no such limitations.

I wish I could say that this esthetically pleasing technique was the result of intelligent planning and deep thought on my part. It was not. Before coming to Cold Spring Harbor, I had studied the movement of cell membranes by placing a few gold particles on

Guenter Albrecht-Buehler, 1975. *(Courtesy of Cold Spring Harbor Laboratory Archives.)*

their dorsal surface. By selecting three of the particles, I could quantify the membrane movement by measuring over time the changes in size and shape of the triangle spanned between the particles.[1] After arriving at the Lab, I thought it would be fascinating to do the same with the ventral surface of cells. But, there was a difficulty: How could I get three particles to stick to the underside of the cells? A truly ridiculous idea occurred to me. If I would cover the glass surface first with gold particles and then plate the cells on top, maybe some particles would stick to the underside and I could watch their movements. So, I did. Of course, the cells stuck no particles to their underside. Instead, they made beautiful tracks.

When I showed the tracks to Jim Watson, I encountered for the first time the incredible speed with which he recognizes the most profound implication of an experimental result. So far, Jim had paid little attention to my work because the first spectacular pictures of the actin architecture of cells had just been generated by Elias Lazarides and others at the Lab.[2] They raised the specter of a direct linkage between biochemical specificity, cell function, and cell structure which, it seemed to me at the time, was foremost on Jim's mind. Still, on one of his rounds through the labs, he came by and asked for an update of what I was doing. I showed him the tracks in the microscope. He was visibly impressed by their esthetics, but quickly returned to reality. "So, what do you do with it?" he demanded. I moved the stage to one of the symmetrical patterns of a sister pair. "Sisters make mirror symmetrical tracks." I said. Jim looked at them for a few seconds, then made some kind of a snorting sound, stood up, and almost stormed out the door without saying anything to me. While I was still puzzling how I could have offended him, Jim McDougall[3] came in. "I just saw Jim. He told me that you have proof that cell migration is programmed," he said.

Notes and References

1. Albrecht-Buehler G. and Yarnell M.M. 1973. A quantitation of movement of marker particles in the plasma membrane of 3T3 mouse fibroblasts. *Exp. Cell Res.* **78:** 59–66.

2. Lazarides E. and Weber K. 1974. Actin antibody: The specific visualization of actin filaments in non-muscle cells. *Proc. Natl. Acad. Sci.* **71:** 2268–2272.

3. Jim McDougall, an expert in transformation of cells by adenoviruses and herpesviruses, arrived at Cold Spring Harbor in 1975 from Birmingham, England. After three years as Head of the Mammalian Cell Genetics laboratory, Jim left Cold Spring Harbor to take up a position as a founding member of the Basic Sciences Division of the Fred Hutchinson Cancer Research Center, Seattle.

DAUGHTER 3T3 CELLS
Are They Mirror Images of Each Other?

GUENTER ALBRECHT-BUEHLER

From the Cold Spring Harbor Laboratory, Cold Spring Harbor, New York 11724

ABSTRACT

Using a new technique to visualize the tracks of moving 3T3 cells and combining it with the visualization of actin-containing microfilament bundles by indirect immunofluorescence (Lazarides, E. and K. Weber. 1974. *Proc. Natl. Acad. Sci. U. S. A.* 71:2268–2272), I present experiments which suggest that: (*a*) 30–40% of the pairs of daughter 3T3 mouse fibroblasts in noncloned cultures have mirror symmetrical actin-bundle patterns. (*b*) The angle between separating daughter cells is approx. 90° or 180° and seems related to the directions of certain actin-containing bundles. (*c*) Approximately 40% of separately moving daughter cells which did not collide with any other cell in the culture performed directional changes in a mirror symmetrical way. Both daughter cells entered the next mitosis at approximately the same time.

I suggest that the actin-bundle pattern, the angle of separation, major directional changes during interphase, and the time of the next mitosis are predetermined by the parental cell.

Considering only the contents of two daughter cells produced by normal mitosis in an established cell line, one would expect them to be identical. This identity in content, however, does not make the cells identical twins. Only after also considering the spatial arrangement of the contents may we identify them as identical twins, as mirror images of each other, or as two randomly packed assemblies of those contents.

Judging by the perfect mirror symmetry of the mitotic spindle, one may suspect that during mitosis other cellular components are likewise mirror-symmetrically distributed between the two compartments which are to become the daughter cells (Fig. 1). Such perfection of mirror symmetry in cells which will soon initiate locomotion and pronounced shape changes can hardly be expected to persist for long.

Yet, if not the perfection of symmetry, at least the left- or right-handedness of the arrangements of the cellular components may persist for a considerable length of time after mitosis. A left-handed three-dimensional cellular arrangement cannot become right-handed by displacements, rotations, or continuous deformations of the cell as a whole. Such operations would merely make it more difficult for the observer to recognize the topological relationship between the cells. The "handedness" can only be destroyed by randomization or the duplication of all components which occurs in cells preparing themselves for the next mitosis. A change to the opposite handedness would require complex individual movements of certain components. In the following, I will use the term "mirror symmetry between daughter cells" in the more general sense of the simultaneous occurrence of one left- and one right-handed cell after mitosis.

If the mirror symmetry between daughter cells would indeed persist for a considerable portion of

DAFNA BAR-SAGI
JAMES R. FERAMISCO

Bar-Sagi D. and Feramisco J.R. 1985. **Microinjection of the *ras* oncogene protein into PC12 cells induces morphological differentiation.** *Cell* **42**: 841–848. (Reprinted, with permission, from Elsevier ©1985.)

THE EARLY 1980S WERE AN EXCITING TIME at Cold Spring Harbor. Cancer research was abuzz with the news about the discovery of the first human oncogene and that its sequences were similar to those of a viral oncogene, *ras*.[1,2,3] Could these proto-oncogenes be converted into true oncogenes by mutation? At Cold Spring Harbor, Mike Wigler's laboratory[1] had shown that the human *ras-1* gene cloned from bladder cancer cells induced transformation of cultured cells, whereas the same gene cloned from normal human cells did not. The only difference between the transforming and normal versions of the *ras* gene was a single-base mutation that caused a change in the amino acid sequence of the *ras* protein.[4]

Our laboratory was keenly interested in helping to unravel the mechanisms by which the activated *ras* oncogene contributed to the cancerous state. We had developed new approaches to studying the cellular activities of *ras* by microinjecting the purified protein into cultured cells and scoring the resulting changes in morphology and cell growth. Microinjection had been brought to Cold Spring Harbor some years earlier by Adolf Graessmann[5] of the Freie University in Berlin, who had taught the technique to one of us (J.R.F.). With several members of our group, and in collaboration with colleagues at the then Smith Kline pharmaceutical company (and

DAFNA BAR-SAGI was a postdoctoral fellow in the Cell Biology group at Cold Spring Harbor between 1984 and 1987 and is now Professor of Biochemistry at New York University Medical School, having previously held a similar position in the Department of Molecular Genetics and Microbiology at the State University of New York at Stony Brook. Her major research interest is the roles of Ras proteins in cell proliferation and oncogenic transformation. dafna.bar-sagi@med.nyu.edu

JAMES R. FERAMISCO'S undergraduate education and graduate training was at the University of California, Davis. He moved to Cold Spring Harbor in 1978 as postdoctoral fellow in the Cell Biology group and was appointed to the staff of the Laboratory in 1978. He is currently a Professor of Medicine and Pharmacology at the University of California, San Diego, working on the signal transduction pathways that regulate cell growth. jferamisco@ucsd.edu

others independently), we showed that microinjection of the activated human Ras protein into quiescent mammalian fibroblasts transiently stimulated cell growth in the absence of growth factors.[6] It appeared that the activated Ras protein could temporarily bypass the growth restraints imposed on normal cells when factors in the medium became limiting. In keeping with these results, Stacey and colleagues had shown that suppression of Ras function by microinjection of antibodies inhibited serum-stimulated growth of cultured fibroblasts.[7] These experiments confirmed the pivotal role of Ras in the intracellular signaling pathways that control cell division. Through microinjection, we now had exactly the assay we needed to explore more fully how Ras worked in living cells and so we were able to carry out experiments in cell systems that, at the time, were not amenable to genetic manipulation. But we never could have predicted what the next series of experiments would show: that the very same Ras oncogene protein that stimulated cell growth was also capable of triggering a biological response that was thought to be the complete opposite of cell proliferation, namely, cell differentiation.

One of us (D.B.S.) had an extensive background in the regulation of cell differentiation. Knowing that exit from the cell division cycle was required for differentiation processes to occur, we wanted to use microinjection to ask whether the Ras oncogene protein would affect (negatively, we guessed) the course of differentiation. To answer the question, we settled on an experiment to find out whether or not the Ras protein could overcome terminal differentiation induced by nerve growth factor (NGF). We needed to use a cell type that would stop growing after NGF was added to the medium and instead would display readily identifiable differentiation characteristics. We chose cells of the pheochromocytoma cell line PC-12, which replicates endlessly in the presence of serum but, upon treatment with the neurotrophic factor NGF, withdraws from the cell cycle and develops neurite-like processes.

The experiment, of course, did not go so easily. As neither of us had ever cultured PC-12 cells, we obtained several samples from nearby biochemists and molecular biologists. As we began to study the cellular responses of the cultivars, we immediately faced a setback. We found that most samples did indeed grow in the presence of serum and show neurite outgrowth in response to NGF. However, all the samples showed substantial spontaneous neurite formation in the absence of NGF and/or incomplete neurite outgrowth when challenged with NGF. We concluded that PC-12 cells were "finicky," as those who studied them had forewarned us. Fortunately, the discoverer of PC-12 cells, Lloyd Greene, had his laboratory in a nearby institution and a short trip with an ice bucket to Manhattan ended with a vial of the cells that turned out to behave perfectly.[8]

In our first experiment, D.B.S. injected PC-12 cells with Ras protein in the presence of normal serum factors, which impeded differentiation. Following overnight incubation, the plan was to add NGF and see if Ras could block the

Jim Feramisco and Dafna Bar-Sagi, 1987. *(Photograph by Susan Lauter, courtesy of Cold Spring Harbor Laboratory Archives.)*

expected neurite outgrowth. Much to our surprise, we found the next day that the cells injected with Ras, and only those, had already sprouted neurites—no NGF was necessary! We spent the next few months developing and expanding this observation to the point where it could be published as the paper that is reproduced on the enclosed DVD.

The observation was not free of strange occurrences. In her initial experiment, D.B.S. left the injected PC-12 cells with their newly discovered Ras-induced neurites on the microscope stage while she called J.R.F. for a look. This took only a few minutes, but when we returned, we saw no neurites. Thinking she was mistaken, we returned the cells to the incubator and looked again in an hour; the neurites had returned, although again, only in the Ras-injected cells. With time, we have learned that the reason for the mysterious disappearance of the Ras-induced neurites is their high sensitivity to changes in temperature and pH. Naturally, because of the totally unexpected nature of this result, we spent months repeating, controlling, probing, and then repeating again the experiments before concluding they were correct.

One personal aside relates to our assumption that this result was so far afield and contrary to the established dogma that growth signals are distinct from and oppose differentiation signals, we felt confident that we would be the only ones in the world with such a result. Well, not long after we submitted our paper,[9] we found ourselves reading a paper published by Noda and colleagues[10] that reached very similar conclusions, only in their experiments, they studied the actions of the Src oncogene rather than Ras. As we still tell each other today, at least there was consolation that another group had reached the same conclusion!

Now, 20-odd years later, we know that the multiple interacting signal transduction pathways regulated by Ras are crucial to normal cellular activities triggered by growth factors of many types. Although not yet fully understood, it appears that the molecular basis of the ability of Ras to induce such disparate responses as cell cycle progression or differentiation lies in the kinetics of activation of a central Ras effector pathway, the ERK cascade (see, e.g., Klesse et al. 1999[11]). In PC-12 cells, transient activation of this cascade promotes cell proliferation, whereas sustained activation induces differentiation. In retrospect, we were fortunate to have chosen to use an oncogenic form of Ras in our experiments. Being constitutively active, it was doing precisely what was needed to drive the differentiation of the cells. Only recently have we learned how cells can distinguish quantitative differences in signal input. As it turns out, the sensing mechanism entails signal-dependent modulation in the levels/activity of transcription factors that turn banks of genes on and off. In 1985, we were excited by the simple result that PC-12 cells extended neurites in response to oncogenic Ras. We never would have guessed that this simple observation would lead us, over the course of two decades, into a complex thicket of intertwined biochemical pathways.

Notes and References

1. Goldfarb M., Shimizu K., Perucho M., and Wigler M. 1982. Isolation and preliminary characterization of a human transforming gene from T24 bladder carcinoma cells. *Nature* **296:** 404–409.

2. Taparowsky E., Suard Y., Fasano O., Shimizu K., Goldfarb M., and Wigler M. 1982. Activation of the T24 bladder carcinoma gene is linked to a single amino acid change. *Nature* **300:** 762–765.

3. Der C.J., Krontiris T.G., and Cooper G.M. 1982. Transforming genes of human bladder and lung carcinoma cell lines are homologous to the *ras* genes of Harvey and Kirsten sarcoma viruses. *Proc. Natl. Acad. Sci.* **79:** 3637–3640.

4. Parada L.F., Tabin C.J., Shih C., and Weinberg R.A. 1982. Human EJ bladder carcinoma oncogene is a homologue of Harvey sarcoma virus *ras* gene. *Proc. Natl. Acad. Sci.* **79:** 3637–3640.

5. Graessmann M. and Graessman A. 1976. "Early" simian-virus-40-specific RNA contains information for tumor antigen formation and chromatin replication. *Proc. Natl. Acad. Sci.* **73:** 366–370.

6. Feramisco J.R., Gross M., Kamata T., Rosenberg M., and Sweet R.W. 1984. Microinjection of the oncogene form of the human H-ras (T-24) protein results in rapid proliferation of quiescent cells. *Cell* **38:** 109–117.

7. Mulcahy L.S., Smith M.R., and Stacey D.W. 1985. Requirement for *ras* proto-onco-

gene function during serum-stimulated growth of NIH 3T3 cells. *Nature* **313**: 241–243.

8. Greene L.A. and Tischler A.S. 1976. Establishment of a noradrenergic clonal line of rat adrenal pheochromocytoma cells which respond to nerve growth factor. *Proc. Natl. Acad. Sci.* **73**: 2424–2428.

9. Bar-Sagi D. and Feramisco J.R. 1985. Microinjection of the *ras* oncogene protein into PC12 cells induces morphological differentiation. *Cell* **42**: 841–848.

10. Noda M., Ko M., Ogura A., Liu D.G., Amano T., Takano T., and Ikawa Y. 1985. Sarcoma viruses carrying *ras* oncogenes induce differentiation-associated properties in a neuronal cell line. *Nature* **318**: 73–75.

11. Klesse L.J., Meyers K.A., Marshall C.J., and Parada L.F. 1999. Nerve growth factor induces survival and differentiation through two distinct signaling cascades in PC12 cells. *Oncogene* **18**: 2055–2068.

Cell, Vol. 42, 841–848, October 1985, Copyright © 1985 by MIT

0092-8674/85/100841-08 $02.00/0

Microinjection of the *ras* Oncogene Protein into PC12 Cells Induces Morphological Differentiation

Dafna Bar-Sagi and James R. Feramisco
Cold Spring Harbor Laboratory
Cold Spring Harbor, New York 11724

Summary

To investigate the possible role of *ras* proteins in the differentiation process signaled by nerve growth factor, we have microinjected the proto-oncogenic and oncogenic (T24) forms of the human H-*ras* protein into living rat pheochromocytoma cells (PC12). PC12 cells, which have the phenotype of replicating chromaffin-like cells under normal growth conditions, respond to nerve growth factor by differentiating into nonreplicating sympathetic neuron-like cells. Microinjection of the *ras* oncogene protein promoted the morphological differentiation of PC12 cells into neuron-like cells. In contrast, microinjection of similar amounts of the proto-oncogene form of the *ras* protein had no apparent effect on PC12 cells. The induction of morphological differentiation by the *ras* oncogene protein occurred in the absence of nerve growth factor, was dependent on protein synthesis, and was accompanied by cessation of cell division. Treatment of PC12 cells with nerve growth factor or cAMP analogue prior to injection did not alter the phenotypic changes induced by the *ras* oncogene protein.

Introduction

The activation of *ras* genes has been implicated in transformation in vitro and tumorigenesis in vivo, but the role of these genes in the sequential events leading to the acquisition of the transformed phenotype is unclear. The *ras*-encoded proteins in mammalian cells are approximately 21,000 daltons (p21), bind guanine nucleotides (Papageorge et al., 1982), and are localized to the inner face of the plasma membrane (Willingham et al., 1980; Shih et al., 1979). Oncogenic *ras* proteins differ from their normal homologs by a single amino acid substitution, usually at position 12 or 61 (Taparowsky et al., 1982; Tabin et al., 1982; Reddy et al., 1982). These mutations do not affect the localization or the nucleotide binding properties of *ras* proteins (Finkel et al., 1984). However, normal p21, encoded by the H-*ras* gene, has been shown to possess an intrinsic GTPase activity that is significantly impaired in the mutated oncogenic protein (Sweet et al., 1984; McGrath et al., 1984; Gibbs et al., 1984). By drawing an analogy between the *ras* proteins and the regulatory proteins (G proteins) of the adenylate cyclase system (Kamata and Feramisco, 1984a; Tanabe et al., 1985; Lochrie et al., 1985; Toda et al., 1985), it has been proposed that the deficiency in GTPase activity of the *ras* oncogene protein could result in the derangement of normal regulatory mechanisms that control cell proliferation (Sweet et al., 1984; McGrath et al., 1984; Gibbs et al., 1984).

Recent studies have indicated a role for the *ras* proteins in growth-factor-mediated control of cell proliferation (Kamata and Feramisco, 1984b; Stacey and Kung, 1984; Feramisco et al., 1984, 1985; Mulcahy et al., 1985). However, growth factors do not affect solely the proliferative capacity of cells. For example, in rat pheochromocytoma cells (PC12), nerve growth factor (NGF) promotes cell differentiation rather than cell proliferation. We have undertaken studies to explore the possible participation of the *ras* proteins in this particular growth-factor-mediated signaling pathway. PC12 cells grow indefinitely in culture as round, chromaffin-like cells. When NGF is added, the cells differentiate into sympathetic neuron-like cells and stop growing (Greene and Tischler, 1976). We show here that in the absence of NGF, microinjection of the human H-*ras* oncogene protein into PC12 cells results in morphological differentiation similar to that induced by NGF. These results indicate that *ras* proteins may exert different biological effects in different cell types, perhaps by utilizing a common molecular mechanism involving the coupling of growth factors to intracellular signals.

Results

Microinjection of the Oncogenic and Proto-oncogenic Forms of the H-*ras* Protein into PC12 Cells

Under routine culture conditions, the pheochromocytoma cells, PC12, proliferate indefinitely as round chromaffin-like cells (Figure 1a). Addition of NGF to the growth medium promotes the morphological differentiation of the PC12 cells into sympathetic neuron-like cells. Within 24 hr the NGF-treated cells flatten and begin to send out short neurite-like processes (Figure 1b), and after several days proliferation ceases and the majority of the cells extend long-branching neurites (Figure 1c). For the microinjection experiments described here, we have utilized the proto-oncogenic and oncogenic forms of the human H-*ras* proteins produced in E. coli (Gross et al., 1985). The proteins have been purified to homogeneity as described previously (Feramisco et al., 1984) and biochemically (Sweet et al., 1984).

Figure 2 shows a PC12 cell that was injected with the H-*ras* oncogene protein (a, b) and a PC12 cell that was injected with the H-*ras* proto-oncogene protein (c, d). In both cases, NGF was not present in the growth medium. As can be seen, the cell that was microinjected with the oncogene form of the H-*ras* protein acquired a polygonal, flattened appearance and extended neuritic processes within 24 hr. In contrast, the cell that was microinjected with similar amounts of the proto-oncogenic form of the H-*ras* protein showed no morphological changes. To identify the cells that had been microinjected with the *ras* proteins, immunofluorescence analysis using anti-*ras* monoclonal antibodies was performed. This analysis was unambiguous because of the low level of endogenous *ras* proteins in normal PC12 cells. As shown in Figure 2, unin-

Harley C.B., Futcher A.B., and Greider C.W. 1990. **Telomeres shorten during ageing of human fibroblasts.** *Nature* **345:** 458–460. (Reprinted, with permission, from Macmillan Publishers Ltd ©1990.)

NEW IDEAS AND DISCOVERIES OFTEN COME from people with very different knowledge and backgrounds talking to each other. The interface between different people's ideas and their scientific biases can challenge dogma and open new possibilities that neither person had previously considered. When I was a Fellow at Cold Spring Harbor Laboratory, I had the freedom to follow interesting new ideas and talk to people whose viewpoints were very different from mine. The accompanying paper on telomere shortening in human cells was the result of such interactions between people in different fields.

As a graduate student at Berkeley with Elizabeth Blackburn from 1983 to 1987, I became fascinated by the mechanisms used to replicate telomeres—the DNA sequences that lie at the very ends of chromosomes. Ten years earlier, Jim Watson had shown that none of the known DNA polymerases were capable of copying the extreme ends of DNA.[1] Jim had used bacteriophage T7 as his example, but the principles he laid down were general and applied to any linear double-stranded DNA. In the succeeding years, many models were proposed to solve the conundrum of end replication.[2] But the first real advance in solving the problem came when Liz Blackburn and Joe Gall established that telomeres of *Tetrahymena* contained simple tandemly repeated sequences.[3] A few years later, yeast telomeres were shown to have a similar structure.[4] Many of the models of end replication were then discarded, leaving just two main possibilities. I set out to test one of these: Namely, that instead of a typical DNA polymerase, a then-unknown enzyme was responsible for elongation of telomeres. This work led to the discovery of telomerase, an enzyme that catalyzes the elongation of telomeres by de novo addition of telomeric repeats.[5]

CAROL GREIDER joined Cold Spring Harbor Laboratory as a Fellow in 1988 and was appointed to the staff of the Lab in the following year, eventually holding the position of Investigator. Carol moved to Johns Hopkins University School of Medicine in 1998, where she is currently Professor and Chair of the Department of Molecular Biology and Genetics and Professor of Oncology. cgreider@jhmi.edu

Carol Greider, 1995. *(Photograph by Marléna Emmons, courtesy of Cold Spring Harbor Laboratory Archives.)*

During my graduate career looking for and then characterizing telomerase, I took several summers off to visit Bruce Futcher[6] who was then an assistant professor at McMaster University. Bruce shared lab space with Calvin Harley, an associate professor interested in the mechanisms of cellular senescence, the term coined by Leonard Hayflick to describe the limited number of doublings primary cells could undergo in cell culture.[7] The cause of this limited replicative capacity was not known. While visiting McMaster in 1985 and 1986, I presented my work on telomerase at joint group meetings with Cal and Bruce's groups. Cal was very interested in telomerase and told me over coffee there had been suggestions that telomeres might play a role in cellular senescence. The link between the two fields was at best speculative and the sequence of the human telomere was not known, so it was not possible to test any specific hypothesis at that stage.

Having identified telomerase in *Tetrahymena* and shown that it contains an essential RNA component, I moved in January 1988 to be a Fellow at Cold Spring Harbor and continued working on telomeres and telomerase. The main focus of my initial work was cloning and characterizing the RNA component of telomerase and studying the enzyme mechanism. However, I continued to talk to Cal, who occasionally visited Bruce Futcher and me and who helped me with mathematical models of enzyme processivity.[8] One day in 1988, Cal's old graduate advisor Sam Goldstein called me to say that he had heard about interesting telomere work from my lab. He asked if Cal had told me about a paper that proposed telomere shortening may trigger cellular senescence. I said that Cal had mentioned it in passing but I had never seen the paper. Then either Sam Goldstein or Cal sent me a paper by Alexis Olovnikov entitled "A theory of marginotomy."[9] Olovnikov had recognized the same problem described by Jim Watson and had proposed that chromosome shortening might be a mechanism to limit the number of divisions of human diploid fibroblasts in culture, the phenomenon first described by Hayflick in 1961.[7] I was startled that I had known nothing of the Olovnikov paper, perhaps because the word telomere is never used in the paper. I called Liz Blackburn and told her about the paper and sent her a copy for her interest. However, even with this potential convergence of telomeres and cellular senescence, it still was not possible to test the idea experimentally because the sequence of the telomeres at the human chromosome ends was not known. There were hints from Howard Cooke's work[10] on the pseudoautosomal region located very near the telomeres on human X and Y chromosomes that the telomeres of human, yeast, and *Tetrahymena* were similar in structure, but the actual DNA sequence was not known.

The public presentation of the human telomere sequence was a classic Cold Spring Harbor story. In May of 1988, the Lab organized a "Genome" meeting. In the early days of the human genome project, before bioinformatics took off, these meetings included sessions on chromosome structure and mechanism. Nick Hastie[11] discussed work that Robin Allshire had done in his lab suggesting that the telomere sequence in humans was related to the *Tetrahymena* TTGGGG repeat sequence. Bob Moyzis[12] presented a compelling talk showing that he had cloned tracts of TTAGGG repeats that hybridized to the ends of human and other mammalian chromosomes. This identification of the human telomere sequence was published soon thereafter in *PNAS*.[13] I already had probes for TTAGGG because this sequence was also present in trypanosomes, which I had worked on previously. Robin Allshire, who was doing a sabbatical as a postdoc at Cold Spring Harbor, was also in the audience and accompanied me immediately to my freezer to get some of the TTAGGG telomere repeat probe.

The day of the telomere session, I called Cal Harley and told him that it was now possible to do the experiments on telomere shortening in human fibroblasts that we had talked about. We designed a blinded set of experiments to examine telomere shortening. Cal sent me DNA from human fibroblasts that had been grown in culture for different numbers of cell doublings. Bruce Futcher blinded the samples and I ran Southern blots of the different samples and probed with the telomere probe. When we developed the gel, sure enough some of the heterogeneous telomere smears were shorter than others. When we unblinded the samples, it was clear that older cells in culture had shorter telomeres. This was very exciting; we then verified the result with several other blinded studies.

We wrote up these results and struggled to get them published in *Nature*. Bruce Futcher taught me the importance of arguing the case when papers of wide interest are initially rejected. The paper finally appeared in May 1990,[14] and a similar paper was published shortly thereafter in an August issue of *Nature* by Robin Allshire and Nick Hastie who analyzed telomeres in cellular senescence and in tumors.[15] Earlier that year, a paper had been published by Titia de Lange and Harold Varmus also showing that telomeres are shorter in human tumor cells than in normal tissue.[16] This convergent data eventually led to the concept of telomere shortening in primary cells in the absence of telomerase and then stabilization of short telomeres in cancer cells by telomerase activation.[17-19]

The ideas that telomere shortening might limit division of normal cells, contribute to aging, and may limit the division of cancer cells were soon picked up by a number of other groups, including Jerry Shay, Woody Wright at Southwestern Medical Centre, Dallas, and Mike West who founded the Geron company, to take advantage of the possible implications of the telomere work. Elegant experiments by a number of different laboratories in subsequent years have confirmed the proposal, first made in 1990, that telomere shortening limits cell division. Cal Harley's group at Geron showed that overexpression of telomerase in primary fibroblasts causes telomeres to elongate and the cells with the elongated telomeres do not enter senescence.[20] We now also know that when telomeres become very short, they no longer function and, instead, signal to the cell that its DNA has been damaged. In cancer cells, this signal can lead to apoptosis[21] or to cell senescence that limits the growth of the tumor.

Most recently, it has become apparent that the level of telomerase in normal cells is tightly regulated and that even a small reduction in telomerase activity in normal cells can cause telomere shortening and cell death. Autosomal dominant dyskeratosis congenita was shown to be caused by mutations in telomerase.[22] The autosomal dominant inheritance is due to haploinsufficiency for telomerase, which causes subsequent telomere shortening that leads to a loss of capacity to renew tissue.[23] Similarly, mutations in telomerase underlie the limited renewal in aplastic anemia and idiopathic pulmonary fibrosis.[24,25] Thus, the proposal made in our 1990 *Nature* paper that telomere shortening might limit replicative capacity of normal cells and may limit cell division of tumors both have turned out to be true. In human disease, we are just beginning to understand the implications of short telomeres. It is already clear that there is limited tissue renewal when telomerase is defective. However, short telomeres may also play a role in age-related disease in individuals who do not carry inherited mutations in telomerase. So far from being over, the telomerase story may still have a long way to run.

Notes and References

1. Watson J.D. 1972. Origin of concatemeric T7 DNA. *Nat. New Biol.* **239:** 197–120.

2. Cavalier-Smith T. 1974. Palindromic base sequences and replication of eukaryote chromosome ends. *Nature* **250:** 467–470.

3. Blackburn E.H. and Gall J.G. 1978. A tandemly repeated sequence at the termini of the extrachromosomal ribosomal RNA genes in *Tetrahymena*. *J. Mol. Biol.* **120:** 33–53.

4. Shampay J., Szostak J.W., and Blackburn E.H. 1984. DNA sequences of telomeres maintained in yeast. *Nature* **310:** 154–157.

5. Greider C.W. and Blackburn E.H. 1985. Identification of a specific telomere terminal transferase activity in *Tetrahymena* extracts. *Cell* **43:** 405–413.

6. Bruce Futcher came to Cold Spring Harbor in 1987 from McMaster University and

left in 2000 for a position at the State University of New York at Stony Brook, where he is now Associate Professor in the Department of Molecular Genetics and Microbiology.

7. Hayflick L. and Moorhead P.S. 1961. The serial cultivation of human diploid cell strains. *Exp. Cell. Res.* **25:** 585–621.

   Hayflick L. 1976. The limited in vitro lifetime of human diploid cell trains. *Exp. Cell Res.* **37:** 614–636.

8. Greider C.W. 1991. Telomerase is processive. *Mol. Cell. Biol.* **11:** 4572–4580.

9. Olovnikov A.M. 1973. A theory of marginotomy. The incomplete copying of template margin in enzymic synthesis of polynucleotides and biological significance of the phenomenon. *J. Theor. Biol.* **41:** 181–190.

10. Cooke H.J., Brown W.A., and Rappold G.A., 1985. Hypervariable telomeric sequences from the human sex chromosomes are pseudoautosomal. *Nature* **317:** 687–692.

11. In 1988, Nick Hastie worked at the MRC Human Genetics Unit at the University of Edinburgh and Robin Allshire was a postdoc in his laboratory. Soon afterward, Allshire moved briefly to Cold Spring Harbor, where he worked on chromosomes in fission yeast and then returned to Edinburgh to study how chromatin structures facilitate chromosomal segregation. Hastie's talk at the 1988 Cold Spring Harbor Genome Mapping and Sequencing meeting was entitled "From yeast chromosomes to human telomeres."

12. Robert Moyzis' talk at the 1988 Cold Spring Harbor Genome Mapping and Sequencing meeting was entitled "Physical maps of human chromosome 16— Status and perspectives."

13. Moyzis R.K., Buckingham J.M., Cram L.S., Dani M., Deaven L.L., Jones M.D., Meyne J., Ratliff R.L, and Wu J.R. 1988. A highly conserved repetitive DNA sequence, $(TTAGGG)_n$, present at the telomeres of human chromosomes. *Proc. Natl. Acad. Sci.* **85:** 6622–6626.

14. Harley C.B., Futcher A.B., and Greider C.W. 1990. Telomeres shorten during ageing of human fibroblasts. *Nature* **345:** 458–460.

15. Hastie N.D., Dempster M., Dunlop M.G., Thompson A., Green D.K., and Allshire R.C. 1990. Telomere reduction in human colorectal carcinoma and with ageing. *Nature* **346:** 866–868.

16. de Lange T., Shiue L., Myers R.M., Cox D.R., Naylor S.L., Killery A.M., and Varmus H.E. 1990. Structure and variability of human chromosome ends. *Mol. Cell. Biol.* **10:** 518–527.

17. Greider C.W. 1990. Telomeres, telomerase and senescence. *BioEssays* **12:** 363–369.

18. Harley C.B. 1991. Telomere loss: Mitotic clock or genetic time bomb? *Mutat. Res.* **256:** 271–282.

19. Harley C.B., Kim N.W., Prowse K.R., Weinrich S.L., Hirsch K.S., West M.D., Bacchetti S., Hirte H.W., Counter C.M., Greider C.W., et al. 1994. Telomerase, cell immortality, and cancer. *Cold Spring Harbor Symp. Quant. Biol.* **59:** 307–315.

20. Bodnar A.G., Ouellette M., Frolkis M., Hold S.E., Chiu C.P., Morin G.B., Harley C.B., Shay J.W., Lichtsteiner S., and Wright W.E. 1998. Extension of life-span by introduction of telomerase into normal human cells. *Science* **279:** 349–353.

21. Qi L., Strong M., Baktiar K., Armanios M., Huso D.L., and Greider C.W. 2003. Short telomeres and ataxia-telangiectasia mutated deficiency cooperatively increase telomere dysfunction and suppress tumorigenesis. *Cancer Res.* **63:** 8188–8196.

22. Vulliamy T., Marrone A., Goldman F., Dearlove A., Bessier M., Mason P.J., and Doka I. 2001. The RNA component of telomerase is mutated in autosomal dominant dyskeratosis congenita. *Nature* **413:** 432–433.

23. Hao L.-Y., Armanios M., Strong M.A., Baktiar K., Feldser D.M., Huso D., and Greider C.W. 2005. Short telomeres, even in the presence of telomerase, limit tissue renewal capacity. *Cell* **123:** 1121–1131.

24. Yamaguchi H., Calado R.T., Ly H., Kajigaya S., Baerlocher G.M., Chanock S.J., Landsdorp P.M., and Young N.S. 2005. Mutations in TERT, the gene for telomerase reverse transcriptase, in aplastic anemia. *N. Engl. J. Med.* **352:** 1413–1424.

25. Armanios M.Y., Chen J.J., Cogan J.D., Alder J.K., Ingersoll R.G., Markin C., Lawson W.E., Xie M., Vulto I., Phillips J.A. 3rd., Lansdorp P.M., Greider C.W., and Loyd J.E. 2007. Telomerase mutations in families with idiopathic pulmonary fibrosis. *N. Engl. J. Med.* **356:** 1317–1326.

LETTERS TO NATURE

even on termini capped by a stretch of non-telomeric DNA, involves recombination, especially if the last few bases of the substrate are able to pair with $C_{1-3}A$ sequence. Alternatively, if telomerase[9-12] exists in *Saccharomyces* and if it can use a substrate in which the telomeric sequences are subterminal, it could mediate those telomere formation events that do not involve recombination.

Our results indicate that telomere formation in yeast is accompanied by telomere–telomere recombination. This recombination requires surprisingly little homology, perhaps because it is a site-specific event promoted by the ability of telomere DNA to assume non-B form structures involving triplex or quadruplex associations[13,14] or G–G[13-15] or C–C base pairing[16]. The junction between telomeric and unique sequence DNA, an apparent hot spot of recombination *in vivo*, also acts as a boundary for Klenow DNA polymerase and S1 nuclease *in vitro*[17,18]. Although the data here and elsewhere[4] do not rule out alternative pathways for telomere formation in yeast, telomere–telomere recombination provides an efficient mechanism for immediate rescue of DNA termini with very short stretches of telomere DNA. In contrast, other events that alter telomere length in yeast do so gradually such that many generations are needed to see a change

in telomere length (reviewed in ref. 19). A working model for telomere formation in yeast by recombination is shown in Fig. 8. This scheme is reminiscent of the bacteriophage T4 'copy choice' model[8,20] that allows complete replication of the T4 genome by a process that requires recombination. As this recombination is non-reciprocal and can result in a net transfer of telomere sequences it could contribute to telomere replication during the normal cell cycle.

If telomere–telomere recombination is important for telomere replication or maintenance, we also expect authentic telomeres to recombine. However, we did not detect transfer of C_4A_4 or C_4A_2 DNA to yeast chromosomes. These results can be explained if yeast telomeres are more likely to recombine with each other than with plasmid termini. This is possible as (1) chromosomal telomeres and internal stretches of $C_{1-3}A$ are more abundant than plasmid termini, and (2) chromosomal telomeres have much greater homology to each other than to C_4A_4 or C_4A_2 DNA. Alternatively, if telomere–telomere recombination is a salvage pathway that acts only on those telomeres with very short stretches of $C_{1-3}A$ DNA, its occurrence at natural chromosomes may be too rare to detect by Southern hybridization.
□

Received 21 December 1989; accepted 16 March 1990.

1. Pluta, A. & Zakian, V. *Nature* **337**, 429–433 (1989).
2. Walmsley, R., Chan, C., Tye, B.-K. & Petes, T. *Nature* **310**, 157–160 (1984).
3. Szostak, J. W. *Nature* **337**, 303–304 (1989).
4. Murray, A., Claus, T. & Szostak, J. *Molec. cell. Biol.* **8**, 4642–4650 (1988).
5. Ahn, B.-Y., Dornfeld, K. J., Fagrelius, T. J. & Livingston, D. M. *Molec. cell. Biol.* **8**, 2442–2448 (1988).
6. Pluta, A., Dani, G., Spear, B. & Zakian, V. *Proc. natn. Acad. Sci. U.S.A.* **81**, 1475–1479 (1984).
7. Bianchi, M. E. & Radding, C. M. *Cell* **35**, 511–520 (1983).
8. Brunier, D., Michel, B. & Ehrlich, S. D. *Cell* **52**, 883–892 (1988).
9. Greider, C. W. & Blackburn, E. H. *Cell* **43**, 405–413 (1985).
10. Zahler, A. & Prescott, D. M. *Nucleic Acids Res.* **16**, 6953–6972 (1988).
11. Shippen-Lentz, D. & Blackburn, E. H. *Molec. cell. Biol.* **9**, 2761–2764 (1989).
12. Morin, G. *Cell* **59**, 521–529 (1989).
13. Sen, D. & Gilbert, W. *Nature* **334**, 364–366 (1988).
14. Williamson, J. R., Raghuraman, M. K. & Cech, T. R. *Cell* **59**, 871–880 (1989).
15. Henderson, E., Hardin, C. C., Walk, S. K., Tinoco, I. Jr & Blackburn, E. H. *Cell* **51**, 899–908 (1987).
16. Lyamichev, V. I. *et al. Nature* **339**, 634–637 (1989).
17. Henderson, E. R. & Blackburn, E. H. *Eukaryotic DNA Replication* Vol. 6 (eds Stillman, B. & Kelley, T.) 453–461 (Cold Spring Harbor Laboratory, New York, 1988).
18. Budarf, M. & Blackburn, E. *Nucleic Acids Res.* **15**, 6273–6292 (1987).
19. Zakian, V. A. *A. Rev. Genet.* **23**, 579–604 (1989).
20. Mosig, G. *A. Rev. Genet.* **21**, 347–371 (1987).
21. Dunn, B., Szauter, P., Pardue, M. L. & Szostak, J. W. *Cell* **39**, 191–201 (1984).
22. Maxam, A. M. & Gilbert, W. *Meth. Enzym.* **65**, 499–560 (1980).
23. Blackburn, E. H. & Gall, J. *J. molec. Biol.* **120**, 33–53 (1978).
24. Henderson, E. R. & Blackburn, E. H. *Molec. cell. Biol.* **9**, 345–348 (1989).
25. Henikoff, S. & Eghtedarzadeh, M. K. *Genetics* **117**, 711–725 (1985).

ACKNOWLEDGEMENTS. We thank G. Smith, A. Taylor, A. MacAuley and the members of our laboratory for comments on the manuscript. This work is supported by the NIH (V.A.Z.) and the Joan Taylor Guild (S.-S.W.).

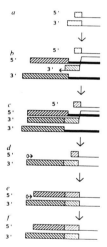

FIG. 3 Model for telomere formation by recombination. Different telomeric repeat sequences (thin-line hatched box) can promote telomere formation in yeast by serving as substrates for the addition of yeast $C_{1-3}A$ repeats (thick-line hatched box). In both cases, the C-rich strand is marked with 5' and the G-rich strand is marked with 3'. Only one terminus of either a linear plasmid or a chromosome is shown. *a*, After replication of the plasmid in yeast and removal of the RNA primer, a single-strand tail is left at the 3' end of the newly replicitied strand (represented by the unpaired thin-line hatched box). *b*, The 3' OH end of the single-strand G-rich strand invades another telomere (donor, represented by thick-line hatched box). At the donor end, most (or all) recombination events are initiated (or resolved) at the junction between telomeric DNA (thick-lined hatched box) and unique DNA (filled-in bar). *c*, The invading G-rich strand is extended by replication using the donor telomere as a template. *d*, After dissociation, the terminus carries an ~300 nucleotide long $G_{1-3}T$ single-strand tail and serves as a template for primase (0) and conventional DNA polymerase mediated replication of the complementary strand (*e*). *f*, Subsequent removal of the RNA primer would still leave a gap at the 5' end of the newly replicated strand, but no sequence information would be lost. (Alternatively, the extended G-rich strand could fold back on itself to form a terminal hairpin and provide the primer for replication of the C-rich strand[15].)

Telomeres shorten during ageing of human fibroblasts

Calvin B. Harley*, **A. Bruce Futcher†** & **Carol W. Greider†**

* Department of Biochemistry, McMaster University, 1200 Main Street West, Hamilton, Ontario L8N 3Z5, Canada
† Cold Spring Harbor Laboratory, Cold Spring Harbor, New York 11724, USA

THE terminus of a DNA helix has been called its Achilles' heel[1]. Thus to prevent possible incomplete replication[2] and instability[3,4] of the termini of linear DNA, eukaryotic chromosomes end in characteristic repetitive DNA sequences within specialized structures called telomeres[5]. In immortal cells, loss of telomeric DNA due to degradation or incomplete replication is apparently balanced by telomere elongation[6-10], which may involve *de novo* synthesis of additional repeats by a novel DNA polymerase called telomerase[11-14]. Such a polymerase has been recently detected in HeLa cells[15]. It has been proposed that the finite doubling capacity

DAVID L. SPECTOR

Spector D.L. 1990. **Higher order nuclear organization: Three-dimensional distribution of small nuclear ribonucleoprotein particles.** *Proc. Natl. Acad. Sci.* **87**: 141–151. (Reprinted, with permission.)

THAT THE CYTOPLASM OF EUKARYOTIC cells is organized into physically identifiable functional units—mitochondria, membranes of several types, lysosomes, ribosomes—has been known for decades. The great advance made by the cell biologists of the 1980s was the discovery that the molecular traffic between and through these organelles is both disciplined and efficient. Looming over all has been the coordinating center, the cell nucleus—huge, challenging, and until recently, by default, regarded as little more than a portmanteau of genes. But to be equal to its tasks, the nucleus too must be organized both spatially and temporally: Banks of genes must be made available at specific stages of the cell cycle or more quickly in response to changes in physiological circumstances; sets of genes that are not immediately required may be silenced epigenetically; RNAs must be processed and delivered efficiently into the cytoplasm through nuclear pores. And the entire machinery needs to be dismantled before, and reassembled after, each cell division.

The roots of the present-day field of nuclear organization draw to a great extent from the elegant early studies (late 1950s to mid 1980s) of Wilhelm Bernhard and his former students (in particular, Guy Blaudin de Thé, Stanislav Fakan, and Edmund Puvion), as well as Karel Smetana, Hewson Swift, and John Sedat—to name just a few. The common goal of these individuals was to relate cellular structure with biological function, by combining electron microscopy (in the early days) and confocal or deconvolution microscopy (more recently) with insight derived from the biochemical studies of others. By the beginning of the 1990s, our understanding of the functional organization of the mammalian cell nucleus and its inner workings was beginning to yield to a powerful fusion of molecular and cellular techniques. And the interest of the broader scientific community quickened as concepts evolved from thinking of the nucleus as an organelle containing nucleoplasmic sap and chromo-

---

DAVID L. SPECTOR (*See biographical footnote on p. 55.*)

somes to an organelle with structurally defined compartments involved in specific functions. However, the composition, geographical distribution, and dwell time of these compartments were unknown.

In the early-to-mid 1980s while a faculty member in the Department of Pharmacology at Baylor College of Medicine, in Houston, I was working on chromosome structure. One morning on his rounds, Harris Busch,[1] who was Chair of Pharmacology, suggested that I have a look at the localization of small nuclear ribonucleoprotein particles (snRNPs), because of their involvement in processing pre-mRNA molecules.[2] This spurred my interest and was the beginning of a long and productive research program aimed at understanding the spatial and temporal aspects of gene expression.

Soon after my arrival at Cold Spring Harbor in 1985, the first generation of commercial confocal microscopes was just becoming available, but our understanding of their capabilities and limitations was not yet realized. Fortunately, we had available a state-of-the-art transmission electron microscope and an integrated image analysis system. We therefore adopted the laborious approach of preparing serial sections of immunolabeled samples for analysis by transmission electron microscopy.

The nuclei of the cells used in the study typically contained between 20 and 50 of these irregularly shaped speckles, which we found to our surprise were connected to each other by a loose reticulum. Although the physical connections between nuclear speckles were likely to have been exaggerated by the enzymatic method used to detect the immunolabeled speckles, the basic findings of the paper still hold: The clusters of small nuclear ribonucleoprotein particles are concentrated in an irregularly shaped pattern of speckles distributed between the surface of the nucleolus and the nuclear envelope, and the locations of the speckles do not coincide with sites of DNA replication or transcription. However, because the *raison(s) d'être* for the speckled pattern remained obscure, we suggested (prophetically as it turned out) that "a new set of organizational criteria that differentiates functional compartments within the nucleus may be needed." For a long time, the model proposed in the paper was extremely controversial. There was fierce debate whether the nuclear speckles were sites at which pre-mRNAs were processed or sites of storage/assembly/modification of pre-mRNA splicing factors. The current thinking is that the latter is correct, as the paper suggests.

Our understanding of the signals involved in the targeting of proteins to nuclear speckles and from nuclear speckles to sites of transcription has advanced significantly since then (for a review, see Lamond and Spector 2003[3]). Recent studies examining residence times have shown that many of the nuclear proteins associated with nuclear speckles as well as other nuclear bodies are remarkably mobile, yet at steady state, nuclear domains are observed (for a review, see Misteli 2001[4]). With further advances in the area of confocal and fluorescence microscopy now on the horizon, it is clear that not only we will be able to address questions relating to the 3D organization of nuclear constituents in living cells over time (4D), but we will also be able to analyze larger sample sizes in significantly less time with significantly higher resolution.

Mona and David Spector, 1986. *(Photograph by Susan Lauter, courtesy of Cold Spring Harbor Laboratory Archives.)*

Not surprisingly, the model of the 3D organization of the mammalian cell nucleus first published in this 1990 manuscript has evolved considerably over the years. At least a dozen distinct domains have been identified in mammalian nuclei, and this number is likely to continue to increase. Later versions of the model have appeared in publications and Web sites dealing with nuclear structure and function (see, e.g., Dellaire et al. 2003[5]; Prasanth and Spector 2005[6]; Spector 2001[7]; Spector 2003[8]).

Notes and References

1. Harris Busch, both a biochemist and a traditional pharmacologist, had a long-time interest in the physical structure of snRNPs and their role in splicing as his laboratory and the laboratory of Sheldon Penman independently discovered snRNAs. The early work of his laboratory is discussed in a review published in 1982: Busch H., Reddy R., Rothblum L., and Choi Y.C. 1982. snRNAs, snRNPs and RNA processing. *Ann. Rev. Biochem.* **51:** 617–654.

2. snRNPs are part of the complex machinery used in the cell nucleus to process newly synthesized pre-mRNA into messenger RNA. snRNPs contain small uridine-rich snRNAs, which catalyze the cleavage of pre-mRNAs. The structural proteins of the snRNPs form a scaffold that holds the active sites of the snRNAs in the optimum configuration for cleavage.

3. Lamond A.I. and Spector D.L. 2003. Nuclear speckles: A model for nuclear organelles. *Nat. Rev. Mol. Cell Biol.* **4:** 605–612.

4. Misteli T. 2001. Protein dynamics: Implications for nuclear architecture and gene expression. *Science* **291:** 843–847.

5. Dellaire G., Farrall R., and Bickmore W.A. 2003. The Nuclear Protein Database (NPD): Sub-nuclear localisation and functional annotation of the nuclear proteome. *Nucleic Acids Res.* **31:** 328–330.

6. Prasanth K.V. and Spector D.L. 2005. The cell nucleus. In *Encyclopedia of life sciences.* John Wiley and Sons, Ltd., Chichester, United Kingdom. doi: 10.1038/npg.eds.0001337.

7. Spector D.L. 2001. Nuclear bodies. *J. Cell Sci.* **114:** 2891–2893.

8. Spector D.L. 2003. The dynamics of chromosome organization and gene regulation. *Ann. Rev. Biochem.* **72:** 573–608.

Proc. Natl. Acad. Sci. USA
Vol. 87, pp. 147–151, January 1990
Cell Biology

Higher order nuclear organization: Three-dimensional distribution of small nuclear ribonucleoprotein particles

(nuclear structure/RNA processing)

DAVID L. SPECTOR

Cold Spring Harbor Laboratory, Cold Spring Harbor, NY 11724

Communicated by James D. Watson, September 15, 1989

ABSTRACT   The structural and functional organization of the cell nucleus has been investigated using three-dimensional reconstruction, immunoelectron microscopy, and high-resolution *in situ* autoradiography. Nuclear regions enriched in small nuclear ribonucleoprotein particles (snRNPs) form a reticular network within the nucleoplasm that extends between the nucleolar surface and the nuclear envelope. The snRNPs occupy ≈18% of the volume of CHOC 400 cell nuclei. The *in situ* sites of DNA replication and transcription are complementary to, rather than coincident with, the nuclear regions concentrated in snRNPs. Based on these data a three-dimensional model of the organization of the mammalian cell nucleus is presented.

While the cell nucleus serves as the repository of the cell's genome, little information is available about the spatial organization of functional components within this organelle. During the interphase portion of the cell cycle, transcription of genes is turned on and off at precise times, DNA is replicated and pre-messenger ribonucleic acid (pre-mRNA) molecules, coding for specific proteins, are processed and transported through the nucleoplasm to the nuclear pores where they are extruded into the cytoplasm as messenger ribonucleoprotein particles. However, relatively little is known about the precise nuclear locations in which these events take place.

Intracellular spatial organization within biological systems is a central factor influencing function at various levels. A striking example of spatial organization is seen in the syncytial blastoderm stages of *Drosophila* embryogenesis (1, 2). During this stage of development, the *Drosophila* embryo is a single cell containing hundreds of nuclei. It has been shown that the spatial distribution of nuclei influences their subsequent cellularization and developmental fate. Furthermore, *Drosophila* polytene chromosomes have been differentially stained with vital dyes and in conjunction with optical sectioning methods it has been shown that these chromosomes are closely associated with the inner surface of the nuclear membrane and contact the membrane at specific sites (3–6). The cytoplasm represents a strong precedent where processes are organized within membrane-bounded compartments such as mitochondria, lysosomes, the Golgi apparatus, and the endoplasmic reticulum. At the cytoplasmic organelle level there is evidence suggesting that cytoplasmic microtubules are involved in determining cell polarity (7), influencing the distribution of other cytoskeletal components (microfilaments, intermediate filaments), and generally in organizing the cytoplasm (8–11). Although the nucleus has no membrane-bounded compartments, it may also be organized into functional domains that have for the most part heretofore been undefinable.

A new set of organizational criteria that differentiates functional compartments within the nucleus may be needed. One of the most noteworthy examples of the relationship between spatial organization and cell function is represented by the nucleolus, which is a distinct biochemical and structural entity within which ribosomal genes and their products are sequestered from the rest of the genome and nucleoplasm. Within this highly specialized non-membrane-bounded region of the nucleus ribosomal gene transcription, ribosomal RNA processing, and preribosomal particle formation occur (12). However, analogous nuclear compartmentalization of events or processes involving transcription by RNA polymerases II or III or processing of their transcripts has not yet been demonstrated. Recently, experimental evidence for a nonrandom distribution of mRNA molecules in the cytoplasm has been presented (13). In the present study I have used specific probes and three-dimensional (3-D) reconstruction techniques to identify subnuclear regions associated with transcription, DNA replication, and pre-mRNA processing. Based on these results I propose a model for the structural and functional organization of the mammalian cell nucleus.

MATERIALS AND METHODS

Cell Culture. CHOC 400 cells obtained from Nicholas Heintz (University of Vermont College of Medicine) were grown on glass coverslips or in 35-mm-diameter Petri dishes in Dulbecco's modified Eagle's medium (GIBCO)/10% fetal bovine serum.

Immunofluorescence. Cells were prepared for immunofluorescence microscopy by published procedures (14). Anti-Sm primary antibody (14, 15) was used at a dilution of 1:250 and fluorescein isothiocyanate conjugated goat anti-mouse IgG (Cappel Laboratories) was used at a dilution of 1:30 for 1 hr at 20°C.

Immunoelectron Microscopy. Cells were prepared for immunoelectron microscopy by previously published procedures (14). Samples were examined in a Hitachi H-7000 transmission electron microscope operated at 60 or 125 kV.

3-D Reconstructions. For 3-D reconstruction cells were incubated with 3-μm-diameter lectin-coated (Con A) polystyrene spheres for 16 hr, fixed, and prepared for immunoelectron microscopy (14). Serial sections were collected on Formvar-coated slot grids and photographs were taken at 60 kV. Data were analyzed using a 3-D reconstruction program with a Kontron image analysis system (IBAS) (Zeiss). Cells that took up three to six spheres were selected for reconstruction. The centers of the spheres in each section were used to align the serial section micrographs before data were entered into the computer program. Data were interactively entered into the system by contouring structures on photographic prints with the use of a mouse. For each image the

---

Abbreviations: 3-D, three dimensional; pre-mRNA, pre-messenger ribonucleic acid; snRNP, small nuclear ribonucleoprotein particle.

TUMOR VIRUSES

The Early Days of Tumor Virus Research at Cold Spring Harbor

JOSEPH F. SAMBROOK

F ROM THE MOMENT IN THE SUMMER of 1967 when Jim was invited to take over as Director of Cold Spring Harbor, it was inevitable that the Laboratory, after decades of work on bacterial and plant systems, would expand into oncogenic animal viruses. Not that prokaryotes were completely mined out—far from it. But, by the late 1960s, a great deal was already known about gene organization and control in *Escherichia coli* and its bacteriophages. The genetic circuitry of prokaryotes had turned out to be so elegant and robust and many people believed it to be a natural—and perhaps exact—model for gene expression in eukaryotic cells. Testing this idea experimentally, however, was a challenge. Mammalian cells were genetically intractable, and growing them in tissue culture was in those days more of an art than a science. Animal viruses seemed to offer the best hope of finding out how mammalian genes were organized and expressed. However, virology at the time was still largely clinically based, and very few people in the field thought in molecular terms.

I happened to be visiting Cold Spring Harbor during those days of crisis in 1967 when the Laboratory's future was in the balance. I was on my way to a Gordon Conference on animal viruses and had come to visit John Cairns, a friend and mentor from my graduate student days in Australia. I had expected to stay with John and Elfie at Airslie. Instead, I found myself sleeping on a very small and very

JOE SAMBROOK *(See biographical footnote on p. 37.)*

hard sofa in the Williams apartment of Rick Davern, an Australian recruited to Cold Spring Harbor by John. Rick told me that John would be tied up for a day or two, that great changes were astir, and that John would perhaps be able to tell me more.

I went to dinner with the Cairns a couple of days later. John seemed to be in a buoyant mood but avoided discussing Cold Spring Harbor except to say that he thought it possible that the Laboratory might soon begin working on the molecular biology of eukaryotic viruses. I was 12 months into a postdoctoral fellowship at the Laboratory of Molecular Biology, and although work was going very well, I was less than happy with life in Cambridge. I told John that I had been thinking about cutting short my stay in England and taking a second postdoctoral position, perhaps in an animal virus laboratory in the United States. John immediately suggested Renato Dulbecco at the Salk Institute and after dinner called him to sound him out. Within 24 hours, I found myself on a plane to California to interview and within 3 months had moved to La Jolla.

About a year later, I was working in the lab when Renato came to my bench and said that Jim Watson was in his office and wanted to see me. Jim immediately started talking about plans to start work on DNA tumor viruses at Cold Spring Harbor and asked in his oblique way whether I might be interested in a job. He suggested that I might want to come to Cold Spring Harbor to help teach the 1968 course on animal viruses, which he felt might benefit from an injection of molecular biology. So, the following summer, my family and I lived for 3 weeks in a moldy and delapidated apartment on the Lab grounds in Williams House, enjoying ourselves immensely with the crowd of graduate students that Jim had brought from Harvard for the summer. You have to remember that this was in the late 1960s when love was free and drugs soft and plentiful. At the end of the course, I received a formal offer from Lab, which I accepted with alacrity.

In retrospect, Jim's choice to begin work on the small DNA tumor viruses was not surprising. Working with John Littlefield in the late 1950s, he had published[1] a description of the anomalous properties of papillomavirus DNA, which were later explained when Jerry Vinograd's group at the Californina Institute of Technology showed that covalently closed circular DNAs could assume a superhelical configuration.[2] Dulbecco's lab, for historical reasons, had always concentrated on polyomavirus, a smaller relative of papillomavirus. But there were good technical reasons for Cold Spring Harbor to switch to SV40, which like polyomavirus could transform cells in culture and could induce tumors when injected into laboratory animals. The best guess at the time was that the small genome of the virus—about 5000 base pairs—was sufficient to code for only six to eight proteins, at least one of which must be an oncogene. Jim believed that identifying and characterizing this gene would open the way to studying human cancer at the molecular level.

Jim's decision was not without risk. The Laboratory was in financial strife, and apart from Demerec Laboratory, which was almost fully occupied, there was no building on the Lab grounds that was functional year-round and, of course, none that was equipped to deal with animal viruses and cells. It was pretty clear that work on animal viruses depended on securing a grant from the National Institutes of Health that would pay for the renovation and re-equipping of James Laboratory and would totally support the salaries and running costs of a research team. The grant was written in late 1968 and funding began in 1969.

In late May 1969, my then wife and I loaded our Toyota with our three small children and all our possessions and drove across the country. We were drawn to Cold Spring Harbor, of course, by the beauty of the place. But also because in summers, Cold Spring Harbor felt like the center of the scientific world, while the isolated winters offered the freedom to work undisturbed. Jim believed that the best way to get good science out of people was to remove as many distractions from their lives as possible. But he also believed that scientists were over the hill by the time they turned 30, which was exactly my age when the Toyota turned into Bungtown Road. I felt that my stay in Cold Spring Harbor might be brief.

The renovation of James Laboratory was well under way but was still far from complete, and it was not to become fully functional until the adjacent offices were completed in 1971—the year that Nixon declared "War on Cancer" with the enactment of the National Cancer Act. That Cold Spring Harbor was in a position to compete successfully for a share of the resulting flood of money was due in large part to Jim's recruiting skills. He enlisted two Harvard graduate students who spent months at Cold Spring Harbor helping to unpack equipment and to set up glassware washing, tissue culture, and biochemistry labs in James. Not surprisingly, they soon realized that this kind of work was not going to earn them a Ph.D., and after a few months, they retreated to Cambridge. More durable were Bill Sugden and Brad Ozanne, who needed to find exemptions from the military draft to avoid being sent to Vietnam. At that time, exemptions from military service could be granted to people who worked at jobs—such as cancer research—deemed by their local draft boards to be in the national interest. Jim's strong support was sufficient to provide Bill and Brad with the immunity they needed to complete research projects in James Laboratory for their Ph.D.s, Bill's from Columbia University and Brad's from NYU. Jim had also arranged for Bernhard Hirt, a skilled tumor virus researcher from the Institute for Cancer Research at the University of Lausanne in Switzerland to take a sabbatical leave in James. Fortunately, we were also able to entice Carel Mulder and Henry Westphal—senior postdoctoral fellows from Dulbecco's group—to join Cold Spring Harbor. It was this nucleus of experienced people that enabled funding of our Cancer Center Grant in 1972. As I remember it, the grant application was a bit of a dog's breakfast with sections on mapping of viral genomes, transcription of SV40, properties of transformed cells, analysis of viral proteins in infected and transformed cells, and electron microscopy of viral DNAs. Mundane and boring now, of course, but cutting edge at the time.

Without this Cancer Center Grant, which with successive renewals was to support work on tumor viruses in both James and Demerec for the next quarter century, the subsequent history of Cold Spring Harbor would have been very different. Because of it, instead of remaining static, we were able to make three crucial appointments in James: Terri Grodzicker, a card-carrying geneticist, came to Cold Spring Harbor after working for several years in Jonathan Beckwith's laboratory at Harvard Medical School; Walter Keller, a German enzymologist and biochemist, came from the National Institutes of Health; and Phil Sharp, a physical chemist, came from Norman Davison's laboratory at the California Institute of Technology. By the beginning of 1972, we were beginning to attract applications from postdoctoral fellows, the most influential of whom was Ulf Pettersson, who brought to us from Uppsala knowledge of human adenoviruses. By this time, molecular analysis of the larger genomes of this group of viruses had been made possible by the arrival of restriction enzymes (see Sharp, p. 223). In those precommercial days, cottage industries were set up in both James and Demerec to produce the range of enzymes required to satisfy the local demand.

The 1974 Cold Spring Harbor Symposium on Tumor Viruses marked a coming of age for the Tumor Virus group. By then, we had produced physical, genetic, and transcriptional maps of the genomes of both SV40[3] (see also Sambrook, p. 95) and adenovirus 2[4] (see also Grodzicker, p. 99). We had identified the region of adenoviral DNA that encoded the viral oncogenes responsible for transformation. We could look ahead with optimism. During the next couple of years, we were able to map the sites of integration of SV40 DNA in the genome of transformed cells (see Botchan, p. 113). Recombinant DNA was on the horizon and the discovery of splicing was just 3 years away (see Chow, p. 105). Elegant work on the biochemistry of the virally encoded tumor antigens (see Tjian, p. 119) and the detailed analysis of the control regions of the SV40 genome were beginning (see Herr, p. 133), as was the use of engineered viruses to express foreign proteins (see Thummel, p. 129).

But the solution to the central question of how the oncogenes of these little viruses could so severely disrupt the growth of cells and cause cancer would remain elusive for another decade—until the illuminating experiments of many people including Earl Ruley (see Ruley, p. 181), Ed Harlow (see Harlow, p. 203), and Doug Hanahan (see Hanahan, p. 193). I once asked Jim whether he had thought that it would take that long to find the answer. "Not really," he said.

Notes and References

1. Watson J.D. and Littlefield J.W. 1960. Some properties of DNA from Shope papilloma virus. *J. Mol. Biol.* **2:** 161–165.

2. Vinograd L., Lebowitz J., Radloff R., Watson R., and Laipis P. 1965. The twisted circular form of polyoma viral DNA. *Proc. Natl. Acad. Sci.* **53:** 1104–1011.

3. Sambrook J.F., Sugden B., Keller W., and Sharp P.A. 1973. Transcription of simian virus 40. III. Mapping of "early" and "late" species of RNA. *Proc. Natl. Acad. Sci.* **70:** 3711–3715.

4. Sharp P.A., Gallimore P.H., and Flint S.J. 1974. Mapping of adenovirus 2 RNA sequences in lytically infected cells and transformed cell lines. *Cold Spring Harbor Symp. Quant. Biol.* **39:** 457–475.

JOSEPH F. SAMBROOK

Sambrook J., Sharp P.A., and Keller W. 1972. **Transcription of simian virus 40. I. Separation of the strands of SV40 DNA and hybridization of the separated strands to RNA extracted from lytically infected and transformed cells.** *J. Mol. Biol.* **70**: 57–71. (Reprinted, with permission, from Elsevier ©1972.)

F ROM ITS INCEPTION IN LATE 1969, a major goal of the Tumor Virus Group at Cold Spring Harbor was to use molecular biology to understand how viruses could impose a malignant phenotype on mammalian cells. We knew that the small DNAs of viruses such as SV40 could encode only a handful of proteins, at least one of which must be a dominant oncogene. If we could find out what this and other SV40-encoded proteins did, perhaps we could establish a portal into the more challenging molecular analyses of human cancers.

Progress was at first painfully slow. The conversion of James Laboratory from a dilapidated home for summer courses to a functional year-round research facility took many months and could not have been achieved without a group of enthusiastic first-year Ph.D. students—some sent from Harvard by Jim Watson and others who were anxious to find a plausible cause to avoid being drafted into the Vietnam War. In those days, involvement in cancer research and a strong letter from Jim was good enough for most draft boards to grant students temporary immunity.

By early 1971, James had begun to fill with newly appointed faculty members and work on SV40 began in earnest. This paper, published in the following year, was an important milestone: It was the first significant publication from the Tumor Virus Group on the molecular analysis of SV40, and it was the first of a score of papers on the transcriptional mapping of DNA tumor viruses that would be published from Cold Spring Harbor in the succeeding five years, culminating in the discovery of splicing.

In the cramped and claustrophobic confines of James, we soon realized that the Laboratory could best be fertile and productive if we worked closely together. The paper, then, was the first product of the intense collaboration and division of labor that was to serve the Tumor Virus Group well for several more years. Walter Keller

---

JOE SAMBROOK (*See biographical footnote on p. 37.*)

generated the RNA drivers used to separate the strands of SV40 DNA, Phil Sharp purified the radiolabeled SV40 DNA, and I set up the hybridization reactions to separate the strands of SV40 DNA and ran the hydroxyapatite columns.

Exactly why we were so pleased to find that the early and late genes mapped to opposite strands of SV40 DNA no longer makes much sense to me. Perhaps it was because the result somehow justified the effort of separating the strands in the first place. In any case, the gratification we felt was short-lived since we had no immediate means to map the early and late RNAs to specific regions of the viral genome. But by the time this paper was published, the first useful restriction enzymes had become available. It was then a simple matter to

Joe Sambrook, early 1970s. *(Photograph by Francoise Kelly, courtesy of Cold Spring Harbor Laboratory Archives.)*

hybridize viral RNAs to the separated strands of specific fragments of the viral DNA and to generate a low-level transcription map of the SV40 genome.[1]

At this stage, thinking about control of eukaryotic gene expression was still dominated by ideas from prokaryotic systems where control was exercised almost exclusively at the transcriptional level by well-characterized operators, promoters, repressors, and sigma factors. Of course, it turned out that the elegant circuits of prokaryotes have little direct relevance to the mechanisms that govern expression of the genes of DNA tumor viruses. But we certainly had a lot of fun standing around the long chalkboard in James trying to talk ourselves into believing that they might.

Notes and References

1. Sambrook J.F., Sugden B., Keller W., and Sharp P.A. 1973. Transcription of simian virus 40. III. Mapping of "early" and "late" species of RNA. *Proc. Natl. Acad. Sci.* **70:** 3711–3715.

J. Mol. Biol. (1972) **70**, 57–71

Transcription of Simian Virus 40

I. Separation of the Strands of SV40 DNA and Hybridization of the Separated Strands to RNA Extracted from Lytically Infected and Transformed Cells

Joe Sambrook, Phillip A. Sharp and Walter Keller

Cold Spring Harbor Laboratory
Cold Spring Harbor, N.Y. 11724, U.S.A.

(*Received 7 April 1972*)

Asymmetric RNA was synthesized *in vitro* from SV40 component I DNA using *Escherichia coli* DNA-dependent RNA polymerase. When denatured, unit-length, single-stranded SV40 DNA was incubated in the presence of 6- to 20-fold excess of asymmetric RNA, about 50% of the DNA (E-DNA) formed DNA–RNA hybrids. The unhybridized DNA (L-DNA) was separated from the DNA–RNA hybrids by chromatography on hydroxyapatite. E-DNA and L-DNA were shown to be the complementary strands of SV40 DNA. After further purification and shearing, the separated strands were hybridized to RNA extracted at different stages of lytic infection and to RNA from transformed cells. "Early" RNA contained sequences complementary to 30% of E-strand DNA; "late" RNA bound to 30 to 35% of E-strand DNA and to 70% of L-strand DNA. RNA from SV3T3 cells hybridized with 50% of E-strand DNA and 15 to 20% of L-strand DNA.

1. Introduction

The pattern of transcription of the DNA of the oncogenic virus SV40 is complicated. In productive infection, some viral RNA sequences appear early and are synthesized throughout the virus growth cycle; others are detected only after the onset of viral DNA synthesis (Oda & Dulbecco, 1968; Aloni, Winocour & Sachs, 1968; Sauer & Kidwai, 1968; Carp, Sauer & Sokol, 1969; Martin & Axelrod, 1969; Martin, 1970; Tonegawa, Walter, Bernardini & Dulbecco, 1970; Sauer, 1971). At late times after infection, 50% of the viral DNA reacts with saturating amounts of RNA, suggesting that the equivalent of one of the two strands of viral DNA is transcribed during lytic infection (Martin & Axelrod, 1969; Martin, 1970). In transformed cells, which contain the complete viral genome (Koprowski, Jensen & Steplewski, 1967; Watkins & Dulbecco, 1967) in an integrated state (Sambrook, Westphal, Srinivasan & Dulbecco, 1968), the situation is not so clear. Virus-specific RNA has been found in cells transformed by SV40 but there is disagreement about the fraction of the genome that is transcribed. Whilst saturation hybridization experiments have yielded data which indicate that the entire viral genome is represented in RNA sequences (Martin & Axelrod, 1969; Martin, 1970), simultaneous competition hybridization experiments suggest that only part of the genome, mostly that which is expressed early in lytic infection, is transcribed (Oda & Dulbecco, 1968; Aloni *et al.*, 1968; Sauer & Kidwai, 1968; Tonegawa *et al.*, 1970; Sauer, 1971). The reason that this unsatisfactory

TERRI GRODZICKER

Grodzicker T., Williams J., Sharp P., and Sambrook J. 1974. **Physical mapping of temperature-sensitive mutations of adenoviruses.** *Cold Spring Harbor Symp. Quant. Biol.* **39**: 439–446. (Reprinted, with permission, from Cold Spring Harbor Laboratory Press ©1975.)

I CAME TO COLD SPRING HARBOR TO DO THE KIND of molecular genetics with tumor viruses that I had done with *Escherichia coli* as a graduate student and postdoc in Jon Beckwith's laboratory at Harvard Medical School. At the time (1972), I knew about Cold Spring Harbor both from its Symposium, which I had attended as a graduate student, and from the yeast and phage summer courses in which fellow postdocs in Boston had been students. I also knew that Jim Watson, as the new(ish) Director, had started Cold Spring Harbor on a new venture: Research on DNA tumor viruses—whose genomes were the only eukaryotic DNAs that one could isolate and analyze in the era before molecular cloning. When Joe Sambrook, who was the head of James Lab, came to Harvard to give a seminar, I asked him if I could come to Cold Spring Harbor to work and he immediately agreed, a far more informal system than exists today.

The project that particularly attracted me was to isolate nonsense mutations in the SV40 DNA segment of a rather esoteric group of viruses known as the adenovirus 2–SV40 hybrids. Although no such mutants had been isolated in any higher eukaryote, it seemed possible that these viruses with their hybrid genomes might provide the means to do so. The hybrids contained a small SV40 segment of DNA, located in a nonessential region of the adenoviral genome, that, we reasoned, could be mutated without affecting the growth of the virus. The SV40 DNA segment was also nonessential but extended the range of cells in which the virus would grow,

TERRI GRODZICKER has been Assistant Director for Academic Affairs at CSHL since 1986, and Editor of the journal *Genes & Development* since 1989. She received her Ph.D. from Columbia University in 1969 and did her postdoctoral research at Harvard Medical School. Moving to CSHL at the end of 1972, she began working with adenovirus and SV40. About six months after arriving, she accepted a staff position and has been there ever since. During the time that Terri worked as a bench scientist, she also started organizing the annual DNA Tumor Virus Meeting and the Laboratory's Molecular Genetics Courses. grodzick@chsl.edu

thus providing us with a phenotype to recognize mutants. As an additional "bonus," I also went about isolating temperature-sensitive mutants with lesions in the adenovirus-2 portion of the genome.

The biochemical analysis of the nonsense aspects of the project eventually turned into a very fruitful collaboration with Ray Gesteland,[1] who was working on translation of viral RNAs in the Demerec Lab. But the viral mutants also became essential genetic tools that allowed us to construct the first physical maps of mutations and, thus genes, of a eukaryotic DNA. This method of mapping later became the proof of principle for construction of physical maps of mammalian and other large genomes by restriction length polymorphisms.[2] In 1974, James Laboratory was like a scientific commune with work and conversations going on at all hours. Apart from Joe, who had done some genetics years before, I was the only geneticist in James: Everyone else was analyzing viral DNA and transcripts, making use of restriction enzymes. The novel analytical procedures that they were developing were certainly a revelation to me. It was sort of amusing that I was totally confident (or blindly optimistic) that nonsense mutants as well as temperature-sensitive mutants could be obtained, whereas most of the biochemists thought that the project would never work out.

In the end, however, it was the collaboration between geneticists and biochemists that made restriction-fragment-length polymorphism (RFLP) analysis work. Besides Joe, Phil Sharp,[3] and myself, the final participant was Jim Williams[4] who had come on sabbatical from Scotland with his collection of adenovirus-5 temperature-sensitive mutants. There were many conversations among all of us as well as with Ulf Petterson[5] who had introduced adenoviruses to Cold Spring Harbor, and it was during one of these conversations that we developed RFLP analysis. A group of us were sitting around in my office slightly hung over on the morning after a party to welcome Jim and his family. I think it was Joe who asked Jim whether he thought that adenovirus 2 and adenovirus 5 could recombine and, if so, couldn't we use differences in restriction sites to physically map mutations. This all seemed far-fetched at the time since only a few restriction enzymes were available and the sites of cleavage of only one enzyme (*Eco*RI) had been mapped on adenovirus-2 DNA. No maps at all of adenovirus-5 DNA had been constructed. But I was very enthusiastic about the project, perhaps because I was influenced by inspiring conversations with Barbara McClintock who more than 40 years before[6] had correlated genetic crossing over with the visible exchange of parts of chromosomes.

We reasoned that if we crossed one of Jim's adenovirus-5 (Ad5) temperature-sensitive mutants with one of my adenovirus 2–SV40 (Ad2SV40) hybrid temperature-sensitive mutants, the genome of the resulting recombinant must contain wild-type Ad5 DNA in the region of the Ad2 mutation and Ad2 DNA in the region of the Ad5 mutation. By carrying out restriction digest analysis of the DNAs of

Terri Grodzicker, 1980. *(Photograph by Ross Meurer, courtesy of Cold Spring Harbor Laboratory Archives.)*

many wild-type recombinants resulting from many independent crosses, a map could be constructed of the smallest regions of wild-type DNA covering the mutations. This would give us a physical map of the genome showing where the mutations were and where the corresponding genes were located. Of course, we did everything else that we could to classify the mutations, such as complementation analyses to divide them into groups, sorting them into "early" and "late" mutants depending on the stage at which growth was blocked under nonpermissive conditions, and analyzing viral proteins. The SV40 DNA sequence also proved to be a useful physical marker when it was present in a recombinant. But it was the combination of genetic crosses with restriction enzyme analysis that was the key to moving genomic analysis to a new level. It was also very satisfying to really see the DNA bands of Ad5, Ad2, and our recombinants on the gels after all those hours of sitting at the flume hood in the virus room doing mutagenesis and crosses. Of course, it was great to have Jim's company in the virus room during those long days while Phil and Joe were busy constructing restriction maps of the parental Ad2ND1 and Ad5 DNAs and doing the analyses of the DNA of the recombinants. We all worked together to deduce the maps of the recombinants.

Just 6 months after the project started, the results were presented at the 1974 Symposium, the topic of which was Tumor Viruses. I was the one to present the talk and was never as nervous in my life as when rehearsing for it. In fact, my collaborators would visit me after rehearsals to make sure that I did not totally stress out—or run. As often happens, my nervousness disappeared a few minutes into the talk and all went very well, although I had to leave immediately afterward to walk off all that nervous energy outside Vannevar Bush Lecture Hall.

By the end of the year, both Phil and Jim had left Cold Spring Harbor, and Joe and I continued the analysis to build up good RFLP maps of a much larger set of recombinants.[7] By then, it seemed clear that our methods could be used to map larger viruses and other genomes, but I certainly did not consider that something as large as the human genome would or could be done any time soon, or as soon as it was.

I went back to the study of nonsense mutants, analyzing in detail the phenotypes of all of the mutant collection that had been built up. I also began to coorganize the annual DNA Tumor Virus Meeting and to look after the molecular biology courses that were held on the top floor of James Laboratory each summer. I had little idea that these activities would eventually take me away from the lab bench to a new life in administration and editing.

Notes and References

1. Ray Gesteland, a biochemist with a passion for events at the ribosome, was Assistant Director at Cold Spring Harbor from 1967 to 1978. He was instrumental in maintaining the equipoise of the Laboratory at a time when the institution had a new Director, was moving into new areas of research, and was expanding its physical facilities at a rapid rate. For an account of these frenzied years, please see: Gesteland R. 2003. CSHL in transition. In *Inspiring science* (ed. J. Inglis et al.). Cold Spring Harbor Laboratory Press, Cold Spring Harbor, New York.

2. RFLPs are differences in sequences between closely related DNAs that can be detected by digestion with restriction enzymes.

3. Phil Sharp's account of his years at Cold Spring Harbor (1971–1974) is published in *Inspiring Science*. (2003. [ed. J. Inglis et al.]. Cold Spring Harbor Laboratory Press).

4. Jim Williams, an expert in adenoviral genetics, worked for many years at the Institute of Virology, Glasgow, before moving to Carnegie Mellon University, Pittsburgh, Pennsylvania.

5. Before Ulf Petterson came to Cold Spring Harbor in the early 1970s, research in James Laboratory was focused exclusively on SV40. Ulf had been a graduate student at Uppsala, Sweden, with Lennart Phillipson, an expert in adenoviruses. Ulf generously taught people at Cold Spring Harbor everything they needed to know about the growth and assay of adenoviruses.

6. Creighton H.B. and McClintock B. 1931. A correlation of cytological and genetic crossing over in *Zea mays*. *Proc. Natl. Acad. Sci.* **17**: 492–497.

7. Williams J., Grodzicker T., Sharp P., and Sambrook J.F. 1975. Adenovirus recombination: Physical mapping of cross-over events. *Cell* **4**: 113–119.

Physical Mapping of Temperature-sensitive Mutations of Adenoviruses

T. Grodzicker, J. Williams,* P. Sharp and J. Sambrook

*Cold Spring Harbor Laboratory, Cold Spring Harbor, New York, 11724; *M.R.C. Virology Unit, Institute of Virology, Glasgow, Scotland*

Mutants of DNA tumor viruses provide useful materials for analyzing the organization of viral genes and their expression in both the lytic and the transforming cycle. Ideally one would like to arrange for a selection or assay system to obtain mutations in specific regions of the genome; in fact, in some cases it has been possible to take advantage of the biological or physical properties of a virus to isolate specific classes of mutants. Examples of such mutants include the transformed cell-dependent mutants of polyoma (Benjamin 1970), host-range mutants of the adenovirus 2 (Ad2)-simian virus 40 (SV40) hybrid Ad2+ND₁ (Grodzicker et al. 1974), and mutants of SV40 whose genomes are not cleaved by restricting endonuclease *Hpa*II (Mertz et al. this volume). In most cases, however, such specific selection systems do not exist, and it is here that the use of conditional lethal mutants, specifically temperature-sensitive (ts) mutants, have been of great value. In principle, mutations leading to a temperature-sensitive phenotype can occur in any region of the genome, provided the gene involved codes for a protein that can exist in an altered labile form, and indeed, sets of ts mutants exist for a variety of viruses. What one then wants to know is the

location of the sites of the mutations on the genome and their organization relative to one another. In some cases, it has been possible to construct genetic maps, based on recombination frequencies, which give a linear order for the ts mutants (Williams et al., this volume; Brown et al. 1973), but one cannot tell the relation between this order and the actual physical location of the mutations on the genome.

What we have done, in the case of adenoviruses, is to align the genetic and physical maps by locating the positions of temperature-sensitive mutants on the physical map. A key feature of the method is the isolation of recombinants from a cross between ts mutants of *different* adenovirus serotypes, rather than from crosses between mutants of the same serotype. An outline of the method, which can be applied to other systems, is shown in Figure 1. We took advantage of the fact that the genomes of adenovirus type 5 (Ad5) and the nondefective Ad2-SV40 hybrid virus, Ad2+ND₁, are closely enough related so that wild-type recombinants can be readily obtained from crosses of Ad5 ts mutants with Ad2+ND₁ ts mutants. However, these viruses differ enough in DNA sequence that recombinants can

Figure 1. Outline of a method for correlation of the genetic and physical maps of adenoviruses. *(a)* x and y are two serotypes of adenovirus whose DNAs differ in their cleavage patterns with restricting endonucleases. A restriction endonuclease cleaves the DNA of serotype x at 4 specific sites (▼), generating 5 unique fragments. The same enzyme cleaves the DNA of serotype y at 6 sites (▲), generating 7 unique fragments. A ts mutant of serotype x bears a mutation at site M1 and a ts mutant of serotype y carries a mutation at site M2. *(b)* A crossover between the sites of the two mutations, M1 and M2, leads to the formation of a wild-type recombinant. *(c)* The recombinant contains DNA sequences from each of the parental serotypes. It contains serotype x sequences in the region of the serotype y mutation and serotype y sequences in the region of the serotype x mutation. The recombinant illustrated contains the 3 leftmost cleavage sites of serotype x and the 2 right-most cleavage sites of serotype y. Restriction enzyme analysis of the recombinant reveals 6 unique fragments: 3 from serotype x, 2 from serotype y, and a new fragment derived from both x and y.

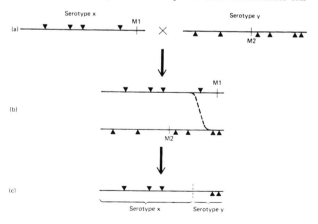

Chow L.T., Gelinas R.E., Broker T.R., and Roberts R.J. 1977. **An amazing sequence arrangement at the 5′ ends of adenovirus 2 messenger RNA.** *Cell* **12:** 1–8. (Reprinted, with permission, from Elsevier ©1977.)

I HAVE NOT LOOKED BACK FOR YEARS ON THE EVENTS in 1976–1977 that led to the discovery of mRNA splicing. The invitation to contribute a reflection to accompany the republication of the landmark paper gives me an opportunity to express some of my thoughts publicly for the first time in nearly three decades.

My husband Tom Broker[1] was recruited in 1975 to Cold Spring Harbor Laboratory as a Staff Scientist and I came along as a postdoctoral fellow. Because of the sudden departure of Hajo Delius,[2] Tom instantly became a Senior Staff Scientist and Lab Chief of the Electron Microscopy (EM) Section upon our arrival in February, 1975. The two of us comprised the EM Section. Jim Watson gave me the freedom to pursue any studies and collaborations that I wished.

Two years out of graduate school at the California Institute of Technology, I was excited by my new research projects on adenoviral transcription, and I also decided to participate in investigations of bacteriophage Mu and *Escherichia coli* transposons, subjects being investigated by the laboratories of Ahmad Bukhari[3] and David Zipser at Cold Spring Harbor. We and Embden, our beloved white boxer, worked and practically lived in the lab. Because the Siemens electron microscope in the EM lab was old and cumbersome, I elected to use the relatively new Philips 200 in the Davenport (now Delbrück) summer course building. I usually had Embden with me during the long hours in the darkened microscope room.

In 1976, I was promoted to Staff Scientist after having a very productive collaboration with Regina Kahmann and Dietmar Kamp, postdoctoral fellows of David

---

LOUISE CHOW, born in Hunan Province, China, completed her undergraduate training in Agricultural Chemistry at National Taiwan University (1965) and carried out her graduate studies in chemistry at the California Institute of Technology (Ph.D., 1973) in the lab of Norman Davidson. After postdoctoral training at the University of California, San Francisco Medical Center, she was appointed Senior Staff investigator in 1977 and Senior Scientist with tenure in 1979. She moved to the University of Rochester School of Medicine in 1984 and subsequently (1993) to the University of Alabama at Birmingham as Professor of Biochemistry and Molecular Genetics. ltchow@uab.edu

Louise Chow and Tom Broker, 1974. *(Courtesy of Cold Spring Harbor Laboratory Archives.)*

Zipser. Together, we discovered that the host range of bacteriophage Mu was determined by G-loop inversion.[4] Jim Watson immediately asked Regina, Dietmar, and me to write an R01 research grant application to the National Institutes of Health (NIH). In March of 1977, I was awarded my first grant as a Principal Investigator.

In 1976, Tom, James M. Roberts (our summer undergraduate research person, or URP, now at Fred Hutchinson Cancer Research Center), and I successfully adapted the R-loop hybridization method developed in the lab of our mentor Norman Davidson[5] to the mapping of adenovirus late mRNAs by EM, the only tool in those days to investigate in detail the organization of such a large genome at the molecular level. In this method, total mRNAs were hybridized to the double-stranded viral DNA genome, generating single-stranded DNA loops opposite the RNA:DNA heteroduplexes. The coding capacities of these mRNAs were determined by the hybridization selection–translation studies of our collaborator James B. Lewis.[6] A manuscript describing the complex arrays of the late RNAs was submitted to *Cell* in February, 1977.[7] In addition to the precise map coordinates of the coding regions for the various late proteins, we described unhybridized tails at both ends of most R-loops. The 3′ branch was certainly due to the poly(A) tail, and conventional wisdom was that the unhybridized 5′ end could be a result of branch migration, but other mechanisms were also deemed possible. We regularly shared our ideas, expertise, and mapping data with Richard Gelinas and Sayeeda Zain, postdocs of Richard J. Roberts in the Nucleic Acid Chemistry Section, and we assisted Gelinas with some of his experiments.

Gelinas identifed a common, capped RNase T1 11-mer derived from two different late mRNAs.[8] In trying to account for this observation, Rich Roberts became obsessed with the idea that the abundant short adenoviral VA RNA served as a primer for late adenoviral mRNA synthesis, analogous to the biosynthesis of influenza virus RNA. Sometime in March, 1977, Roberts learned of a rumor from Joe Sambrook, then in England, that Phil Sharp[9] was onto something big. Sambrook and Roberts speculated that it must be an experiment involving EM as Phil (as had Tom and I) had trained as an electron microscopist in Norman Davidson's lab.

In late March, Roberts and Gelinas proposed a collaborative EM study to test Roberts's pet hypothesis. They suggested that we hybridize restriction fragments to R-loops in full-length adenoviral DNA. We agreed to participate but recommended instead the use of separated strands of restriction fragments to avoid the loss of potentially very short RNA:DNA heteroduplexes to branch migration. Tom and I then devised a hybridization strategy that would optimize the thermodynamic and kinetic conditions favorable to successful annealing and stabilization of the anticipated but possibly very short DNA:RNA heteroduplexes.

The initial DNA fragment chosen was *Hind*III B (map coordinates 17–31) because the VA genes are located near map unit 29. Thus, the predicted heteroduplexes on the 5′ ends of mRNAs in R-loops could easily be recognized. Jim Lewis generously provided late adenoviral mRNA, and Gelinas prepared the separated strands of *Hind*III B. The experiment worked on our very first attempt in the first week of April. But, unexpectedly, the same *two* short regions of the slow-strand DNA hybridized to the 5′ ends of many different late mRNAs in R-loops, creating a reproducible loop in the single-stranded DNA fragment. Tom and I realized immediately that the VA RNA had nothing to do with the biogenesis of late mRNAs because neither complementary region was near the VA genes. Working feverishly with separated strands of additional restriction fragments, by the third week of April, we had determined that all of the late mRNAs encoded by the rightwardly transcribed strand contained the same *three*-part leader at their 5′ ends. The EM data were continuously fed to everyone involved as soon as I had the approximate locations for coding sequences of the 5′ leader RNA by guesstimating the DNA segment lengths in the heteroduplexes as they were visualized in the microscope. After the three genomic sites of origin were pinpointed by EM, Dan Klessig[9] was finally able to protect the capped 11-mer on two late mRNAs with the appropriate restriction fragment.

The EM findings immediately accounted for the common capped oligomer and also for the second-site hybridization of late mRNAs of adenovirus and adeno–SV40 hybrid viruses, an observation consistently made by Jim Lewis, John Atkins, Carl Anderson, and Dan Klessig[10] in Ray Gesteland's Protein Synthesis Section and by Ashley Dunn and John Hassell[11] in James Laboratory. Four manuscripts were submitted simultaneously to *Cell* in early June. Our results also explained why Sayeeda Zain could not locate, in repeated experiments, a sequence consistent with the common capped oligomer at the 5′ end of the fiber gene, a finding on which Roberts had not previously put much weight. Unfortunately, Sayeeda was out of the country during that hectic but extremely exciting and productive period.

After we had determined the origin of the 5′ end of the late mRNAs, Roberts received a manuscript by Berget, Moore, and Sharp, in press in *PNAS*. It described unhybridized tails at both ends of the adenoviral hexon mRNA upon hybridization to a single-strand restriction fragment spanning the coding region. The source of the 5′ noncomplementary sequence was not identified. We were all relieved.

In late April, 1977, Roberts announced our collaborative EM observation in an RNA Symposium organized by Jim Darnell[12] at The Rockefeller University. One of the speakers was Lennart Philipson,[13] then on sabbatical with David Baltimore[14] at the Massachusetts Institute of Technology. When the paper by Berget et al. was published,[15] it had been revised to include an experiment identifying the tripartite leader in heteroduplexes of hexon mRNA and single-stranded DNA from a longer restriction fragment. Baltimore was acknowledged to have suggested the experiment.

Tom presented the findings described in the four *Cell* manuscripts[16] and that of Zain's work in the Cold Spring Harbor Symposium in June 1977 to a surprised and excited audience.[17] Sharp presented results from his laboratory at the Symposium as well.[18] The concept of mRNA splicing gained acceptance quickly because (1) it was clearly shown using multiple, complementary techniques anchored by the EM visualization and (2) it explained anomalous results of several research groups represented in the audience, including labs working on SV40, the β-globin gene, and the ovalbumin gene.

Our EM work not only demonstrated mRNA splicing, but also provided the first description of the novel concepts of eukaryotic gene expression using alternative mRNA splicing and alternative poly (A) sites. However, this latter contribution has generally been overlooked. We subsequently described the use of alternative promoters in the biogenesis of adenoviral early region mRNAs.[19] These closely coupled phenomena greatly expand the addressable genetic content and versatility of eukaryotic genomes.

The award of the 1993 Nobel Prize for Physiology or Medicine to Phil Sharp and Rich Roberts failed to acknowledge the central contributions that Tom and I had made. We were deeply disappointed. Only Susan Gensel (then the Librarian and Public Relations Officer at Cold Spring Harbor Laboratory) called, inviting me to a celebratory bash for Roberts, an event we did not attend. For whatever reasons, we were not included in the 20th anniversary commemoration of splicing held at the Laboratory.

Tom and I were very fortunate to have had an essential role in the discovery of mRNA splicing, having possessed the necessary knowledge and expertise, and being in the right place at the right time. Because I was brought up in Confucian teaching, I found overt self-promotion distasteful. Furthermore, being young and naïve, I believed that the published work would speak for itself and that we would be accorded appropriate recognition without having to campaign for it. Many of us learn sooner or later that this is not necessarily so. Today, proper distribution of credit can still be a problematic issue, and politics continues to be a major factor in many major awards. Regardless, we have moved on and forward because of our love for scientific inquiry and the satisfaction derived from discovery. Since leaving Cold Spring Harbor in 1984, our focus has been on the medically important human papillomaviruses and their associated diseases, a pursuit which has been challenging, successful, and gratifying, as well as broadening our horizons.

Notes and References

1. Tom Broker's Ph.D. thesis in the laboratory of Robert Lehman in Biochemistry at Stanford used electron microscopic and genetic techniques to map DNA recombination pathways and structures in bacteriophage T4. This work was the first one to recognize DNA branch migration during DNA recombination. His postdoctoral research (1972–1975) at Caltech in the lab of Norman Davidson led to the first development of the avidin:biotin complex technique for conjugating macromolecules and was applied to electron microscopic mapping of gene:RNA transcript complexes and protein:DNA complexes. He joined Cold Spring Harbor as Senior Investigator and EM Lab Chief in 1975 and was promoted to Senior Investigator with tenure in 1979.

2. In the late summer of 1969, Hajo Delius joined Cold Spring Harbor Laboratory from the University of Geneva to take charge of a new Siemens electron microscope in the basement of Demerec Laboratory. During his 5 years at Cold Spring Harbor, he collaborated with just about everyone working at the Laboratory. In 1974, Hajo left for a position at the University of Munich.

3. See Harshey, p. 7.

4. Kamp D., Kahmann R., Zipser D., Broker T.R., and Chow L.T. 1978. Inversion of the G DNA segment of phage Mu controls phage infectivity. *Nature* **271**: 577–580.

5. Norman Davidson was a member of Caltech's faculty from 1946 until he died at age 85 in 2002. A physical chemist by training, Norman's goals were to demystify biology by applying physical and chemical principles to problems such as denaturation of DNA and stability of DNA:RNA hybrids. Later, his laboratory developed R-loop mapping as an electron microscopic technique to visualize RNA complexes. It was this technique that was first used by Louise Chow and Tom Broker to show that adenoviral messenger RNAs are spliced.

6. Jim Lewis worked on the molecular biology of tumor viruses, particularly adenoviruses at Cold Spring Harbor Laboratory from 1973 until 1980, first as a postdoctoral researcher and then as a Staff Investigator and Senior Staff Investigator. He continued his research at the Fred Hutchinson Cancer Research Center, Seattle.

7. Chow L.T., Roberts J.M., Lewis J.B., and Broker T.M. 1977. A map of cytoplasmic RNA transcripts from lytic adenovirus type 2, determined by electron microscopy of RNA:DNA hybrids. *Cell* **11**: 819–836.

8. Gelinas R.E. and Roberts R.J. 1977. One predominant 5′-undecanucleotide in adenovirus 2 late messenger RNAs. *Cell* **11**: 533–544.

9. After several years as a senior postdoctoral fellow in Norman Davidson's laboratory at Caltech, Phil Sharp moved to Cold Spring Harbor in 1970 to gain experience in working with animal cells and viruses. In James Laboratory, Phil became involved in many projects, including restriction mapping and identifying and mapping viral RNAs in cells infected or transformed by SV40. For the last year of his stay, he collaborated with other investigators in James on adenoviral transcription, a project that he took with him to MIT in 1974.

10. Dan Klessig was a Ph.D. student and Carl Anderson and John Atkins were postdoctoral fellows in Ray Gesteland's laboratory in Demerec.

11. Ashley Dunn and John Hassell were postdoctoral fellows in Joe Sambrook's laboratory in James.

12. Jim Darnell had shown that cells infected with adenoviruses contained short-lived transcripts that stretched from one end of the genome to the other. At the time, no one could understand the role of these RNAs. Once splicing had been discovered, the significance of Darnell's experiments became apparent: It was immediately obvious that the giant nuclear adenoviral RNAs were the precursors to functional spliced mRNAs.

13. Lennart Phillipson, a Swedish virologist of great distinction, was an early pioneer in the application of molecular biology techniques to the study of adenoviruses. Working in Uppsala, he trained several of the young Swedes who came to the

United States as postdoctoral fellows in the early 1970s. Among these was Ulf Pettersson, who brought much needed knowledge of adenoviruses to James Laboratory. Over the course of Ulf's stay, the Laboratory's focus gradually switched from SV40 to the larger, more complex and more sophisticated adenoviruses.

14. At the time of the events described in this essay, David Baltimore was working at MIT where he was to establish the Whitehead Centre. For some years, Baltimore's own work had been concerned with RNA viruses and retroviruses, but he had retained an interest in DNA viruses from his time in Renato Dulbecco's laboratory in the late 1960s. In 1975, Baltimore shared the Nobel Prize for Physiology or Medicine with Howard Temin and Renato Dulbecco.

15. Berget S.M., Moore C., and Sharp P.A. 1977. Spliced segments at the 5′ terminus of adenovirus 2 late mRNA. *Proc. Natl. Acad. Sci.* **74:** 3171–3175.

16. Chow L.Y., Gelinas R., Broker T.R., and Roberts R.J. 1977. An amazing sequence arrangement at the 5′ ends of adenovirus 2 messenger RNA. *Cell* **12:** 1–8.

    Klessig D.F. 1977. Two adenovirus mRNAs have a common 5′ terminal leader sequence encoded at least 10 kb upstream from their main coding regions. *Cell* **12:** 9–22.

    Dunn A.R. and Hassell J.A. 1977. A novel method to map transcripts; evidence for homology between an adenovirus mRNA and discrete multiple regions of the viral genome. *Cell* **12:** 23–36.

    Lewis J.B., Anderson C.W., and Atkins J.F. 1977. Further mapping of late adenovirus genes by cell-free translation of RNA selected by hybridization to specific DNA fragments. *Cell* **12:** 37–44.

17. Broker T.R., Chow L.T., Dunn A.R., Gelinas R.E., Hassell J.A., Klessig D.F., Lewis J.B., Roberts R.J., and Zain B.S. 1978. Adenovirus-2 messengers—An example of baroque molecular architecture. *Cold Spring Harbor Symp. Quant. Biol.* **42:**(Part 1) 531–553.

18. Berget S.M., Berk A.J., Harrison T., and Sharp P.A. 1978. Spliced segments at the 5′ termini of adenovirus-2 late mRNA: A role for heterogeneous nuclear RNA in mammalian cells. *Cold Spring Harbor Symp. Quant. Biol.* **42:**(Part 1) 523–529.

19. Chow L.T., Broker T.R., and Lewis J.B. 1979. Complex splicing patterns of RNAs from the early regions of adenovirus-2. *J. Mol. Biol.* **134:** 265–303.

Cell, Vol. 12, 1–8, September 1977, Copyright © 1977 by MIT

An Amazing Sequence Arrangement at the 5' Ends of Adenovirus 2 Messenger RNA

Louise T. Chow, Richard E. Gelinas, Thomas R.
Broker and Richard J. Roberts
Cold Spring Harbor Laboratory
Cold Spring Harbor, New York 11724

Summary

The 5' terminal sequences of several adenovirus 2 (Ad2) mRNAs, isolated late in infection, are complementary to sequences within the Ad2 genome which are remote from the DNA from which the main coding sequence of each mRNA is transcribed. This has been observed by forming RNA displacement loops (R loops) between Ad2 DNA and unfractionated polysomal RNA from infected cells. The 5' terminal sequences of mRNAs in R loops, variously located between positions 36 and 92, form complex secondary hybrids with single-stranded DNA from restriction endonuclease fragments containing sequences to the left of position 36 on the Ad2 genome. The structures visualized in the electron microscope show that short sequences coded at map positions 16.6, 19.6 and 26.6 on the R strand are joined to form a leader sequence of 150–200 nucleotides at the 5' end of many late mRNAs. A late mRNA which maps to the left of position 16.6 shows a different pattern of second site hybridization. It contains sequences from 4.9–6.0 linked directly to those from 9.6–10.9. These findings imply a new mechanism for the biosynthesis of Ad2 mRNA in mammalian cells.

Introduction

In contrast to the detailed knowledge of the mechanics of transcription in procaryotic cells (Losick and Chamberlin, 1976), little is known about this process in eucaryotic cells. Several possible schemes exist: one, analogous to the bacterial system, requires independent promoters for each mRNA; a second postulates the production of long primary transcripts in the nucleus which are subsequently cleaved to yield individual mRNAs (Darnell, Jelinek and Molloy, 1973); and a third invokes the use of RNA primers coded at one region on the genome but acting at some other region(s) and becoming elongated into mRNAs (Dickson and Robertson, 1976). Experiments to test these hypotheses directly have been hampered by the complexity of the eucaryotic genome. We have chosen to study these processes in a simpler system—lytic infection of human cells by adenovirus 2 (Ad2).

Ad2 DNA is transcribed by RNA polymerase II (Price and Penman, 1972; Wallace and Kates, 1972), and its transcription shows features charac-

teristic of that of the host genome (Lewin, 1975a, 1975b). For example, long polyadenylated transcripts appear in the nucleus, but only a small percentage of this nuclear RNA appears as polyadenylated mRNA on cytoplasmic polysomes (Philipson et al., 1971). These mRNAs are "capped" at their 5' ends (Moss and Koczot, 1976; Sommer et al., 1976). Gelinas and Roberts (1977) found that most Ad2 mRNAs isolated at late times during infection contain the same "capped" 11 nucleotide sequence at their 5' ends. This sequence was sensitive to ribonuclease cleavage in mRNA:DNA hybrids (Gelinas and Roberts, 1977; Klessig, 1977) and led to the suggestion that this 5' terminal sequence might not be coded immediately adjacent to the main body of the mRNA.

Thomas, White and Davis (1976) have shown that individual RNA molecules can be displayed as RNA displacement loops (R loops) in the electron microscope, and map coordinates have been obtained for many Ad2 mRNAs (Meyer et al., 1977; Chow et al., 1977). In the present studies, we have used mRNAs visualized in such R loops to examine more closely the sequences present at the 5' end of late Ad2 mRNAs.

Results

R loops were formed between Ad2 DNA and polysomal RNA isolated 22 hr after Ad2 infection. The 5' ends of the mRNA should form single-stranded projections if they are not coded immediately adjacent to the rest of the mRNA, and so might be visualized by hybridization to a single-stranded DNA fragment containing their complement. We therefore prepared a set of restriction endonuclease fragments of the Ad2 genome, separated their strands by agarose gel electrophoresis (Hayward, 1972; Sharp, Gallimore and Flint, 1974) and added each single strand in turn as a third hybridization component after the preparation of the R loops. Since R loops were formed from a mixed population of late mRNAs, many different species were examined simultaneously. By using a restriction endonuclease fragment as the single-stranded probe, complicated structures which might arise from hybridization of the probe to the single-stranded DNA segment of the R loop were limited to one region of the genome. Figure 1a shows the results of such an experiment using the slow strand of Hind III-B (map position 17.0–31.5) as the single-stranded DNA probe. The probe hybridized with the 5' end of hexon mRNA in the R loop but not with the displaced DNA strand. It adopted a looped configuration, indicating that sequences from the 5' end of the mRNA were complementary to two separate regions within the probe. The 5' ends of other

Botchan M., Topp W., and Sambrook J. 1976. **The arrangement of simian virus 40 sequences in the DNA of transformed cells.** *Cell* 9: 269-287. (Reprinted, with permission, from Elsevier ©1976.)

————————

AT MY INTERVIEW FOR A POSTDOCTORAL POSITION at Cold Spring Harbor during the winter of 1970, Jim Watson asked me why I was interested in SV40. I was well prepared for the question and without hesitation replied that the SV40-transformed cell provided a very simple system to study a very complex problem. Joe Sambrook and Heiner Westphal, as postdoctoral fellows with Renato Dulbecco, had already proven that in the stably transformed cell, the SV40 DNA had become covalently linked to the cellular chromosome.[1] Since both Heiner and Joe were by then working at Cold Spring Harbor, it seemed that the Laboratory was the place for me. I wanted to understand how the DNA of a small virus might attach itself to cellular chromosomes and maintain itself, essentially indefinitely, in a dividing eukaryote cell and express only a limited set of its genes. A good speculation was that these viral gene products made in this latent state had something to do with maintaining the cancer-like changes in growth and behavior displayed by transformed cells. For me, the problem of integration was interleaved with another—the rescue of integrated SV40 DNA when the transformed cell was fused with a "permissive" monkey cell. If the virus had a way to get out of the chromosome, and once again could resume vegetative replication, it had to be pretty smart to get integrated in the first place. In the early 1970s, the only model for these types of processes was provided by the prophages of prokaryotes, and one naturally thought of repressors and site-specific recombination for the initiation and maintenance of integration and activators or inducers for the reversal.

———————————————————

MIKE BOTCHAN is currently Co-chair and Professor in the Department of Biochemistry & Molecular Biology, University of California, Berkeley. He came to Cold Spring Harbor in 1972 with a Ph.D. in Biophysics from Berkeley and worked on problems ranging from the integration and excision of SV40 DNA to the terminal sequences of bacteriophage Mu DNA. Since returning to Berkeley in 1980, he has published extensively on the molecular biology of papillomaviruses and, more recently, on the core machinery for initiating DNA replication in *Drosophila*. mbotchan@berkeley.edu

Mike Botchan, 1976. *(Courtesy of Cold Spring Harbor Laboratory Archives.)*

At the time, I did not have a specific plan as to how best to push these issues further, but the immediate questions were so clear that it seemed to me only a matter of time before a precise and elegant approach would reveal itself—yes, some arrogance was helpful. The immediate questions were: Did viral DNA integrate at a certain chromosome site(s)? Was attachment via a specific viral DNA sequence necessary for transformation? Was this sort of recombination likely to be a significant pathway of gene exchange between host and virus in a natural setting? In the background was the question of the effects of chromosomal environment on gene expression.

The obvious approach was of course restriction enzyme mapping and DNA:DNA hybridization. After arriving at Cold Spring Harbor, I spent some time convincing myself that restriction enzymes could reproducibly cleave chromosomal DNA to completion[2] and that maps of what we called "single-copy" genes could at least in principle be derived if fragment lengths detected by hybridization could be recorded. So all we needed was a convenient way to fractionate DNA fragments of different lengths, perform the hybridization, and detect the signals. The advent of high-specific-activity radioactive nucleotide precursors (often greater than 10–20 Curies per millimole) and the development of in vitro labeling of DNA by nick translation[3] as taught to me by Tom Maniatis[4] provided us with the means to routinely produce hybridization probes that could detect SV40 DNA at one part in a million in as little as a microgram of total cellular DNA. However, high resolution of DNA fragments was possible only by gel electrophoresis, and I was never able to hybridize probes to fragments embedded in gels or to elute from the gels sufficient amounts of the test DNA. For a while, the problem seemed insoluble. However, the path to generate restriction maps of integrated viral DNA was suddenly cleared when Mike Mathews[5] returned from a holiday visit to England with the news of Ed Southern's method[6] of transferring DNA from gels to nitrocellulose filters.

In the paper reprinted here, we provided detailed descriptions of the organization of viral DNA sequences integrated into the genomes of an isogenic set of 11 rat cell lines transformed by SV40. Although many well-characterized lines of

SV40-transformed cell lines had been described in the literature, it was often impossible to trace their provenance or history. Without this information, we would be unable to solve the potentially confounding problem of the stability of the integrated DNA. Comparing the maps of viral DNA from different cell lines might inadvertently bring us to study of the evolution of the integration position of viral DNA or its flanking cellular sequences at different times in the cells' history, obscuring any initial commonality. For technical reasons, it was also important to use a set of isogenic cell lines for which we could be sure of having viral probes homologous to the transforming virus. Bill Topp[7] had been using a variety of methods to generate transformed cells in order to understand the full spectrum of transformed phenotypes, and he was thus able to provide a set of independent transformants that were perfect for our studies. The data we obtained showed conclusively that the integration patterns were indeed stable and that the prophage model was not applicable. The integration process seemed rather random with regard to how the viral DNA was integrated into the chromosome with one important exception: Maps of the integrated fragments of the viral genome showed that in all cases, at least one intact "early region" capable of expressing full-length SV40 T antigen was present in the chromosomal DNA. No free viral DNA was ever detected. Thus, we speculated that integration was a cellular process that might operate on any foreign DNA: Only by selection for a phenotype could this process show its hand. In retrospect, we now know that the integration of SV40 DNA was the result of cellular mechanisms that had evolved to repair damaged chromosomal DNA and that randomly broken viral DNA was joined together and linked to the chromosome to eliminate free ends. The nonhomologous end-joining pathway is essential in dividing cells because DNA replication forks break at an alarming rate, and without these recombination-driven healing mechanisms, chromosomal stability would be in severe jeopardy.

The paper was important in several ways: First, the work put to rest many of the issues of SV40 integration and excision. We later found that only lines of transformed cells containing integrated head-to-tail tandem copies of the viral genome could yield infectious virus after fusion with permissive simian cells.[8] This result strongly indicated that excision of the viral genome involved homologous recombination and was not simply a reversal of the integration process. The work also identified a cell line (14B) that carried a single integrated copy of the SV40 genome. Revertants of 14B cells that no longer displayed the transformed phenotype turned out to have suffered mutations in the viral T-antigen gene, proving beyond doubt that T antigen was needed for maintenance of transformation. Perhaps the longer-lasting impact of the paper, however, was methodological: The techniques pioneered to detect individual genes within total cellular DNA and the logic employed to generate restriction maps of single-copy sequences are still widely used today.

What remains for me aside from these discoveries are the intense moments of collaboration that made the work exciting. I must end with a few words about my mentor, Joe Sambrook. I loved his critical nature and argumentative style: Perhaps it was the immigrant Coney Island upbringing in my background that resonated with his Liverpool working-class manners. Be that as it may, Joe took immense joy in finding problems with our thinking and strove for what I thought was an extraordinary level of perfection. To the extent that Botchan et al. is as complete and as well crafted as it is owes much to our collaborative efforts.

Notes and References

1. Sambrook J., Westphal H., Srinivasan P.R., and Dulbecco R. 1968. The integrated state of viral DNA in SV40-transformed cells. *Proc. Natl. Acad. Sci.* **60**: 1288–1295.

2. Botchan M., McKenna G., and Sharp P.A. 1974. Cleavage of mouse DNA by a restriction enzyme as a clue to the arrangement of genes. *Cold Spring Harbor Symp. Quant. Biol.* **38**: 383–395.

3. Kelly R.B., Cozzarelli N.R., Deutscher M.P., Lehman I.R., and Kornberg A. 1970. Enzymatic synthesis of deoxyribonucleic acid. XXXII. Replication of duplex deoxyribonucleic acid by polymerase at a single strand break. *J. Biol. Chem.* **245**: 39–45.

4. Tom Maniatis came to Cold Spring Harbor as a refugee from Harvard in the mid 1970s, a time when poorly informed activists of Cambridge were calling for a ban on all research on recombinant DNA. During his year at Cold Spring Harbor, Tom and his host co-workers succeeded in producing the first full-length cDNA clone of a mammalian mRNA: rabbit β-globin.

5. Mike Mathews worked for more than two decades at Cold Spring Harbor—first as an early member of the Tumor Virus Group in James Laboratory and then in Demerec Laboratory as leader of a team that over the years worked on diverse aspects of regulation of gene expression in mammalian cells. Mike Mathews left Cold Spring Harbor in 1996 to take the position of Professor and Chairman of the Department of Biochemistry and Molecular Biology at the Medicine and Dentistry School of New Jersey.

6. Southern E.M. 1975. Detection of specific sequences among DNA fragments separated by gel electrophoresis. *J. Mol. Biol.* **98**: 503–517.

7. Bill Topp, like many other people, came to Cold Spring Harbor as a result of a short conversation with Jim Watson. The pair met in Princeton in 1974, when Bill was finishing his Ph.D. in chemistry with John Hopfield. At Cold Spring Harbor, Bill worked for a year or two in Bob Pollack's Mammalian Cell Genetics Group in Demerec and then moved to James Laboratory as a member of the Tumor Virus Group. Bill left the Laboratory in the mid 1980s to work in the biotechnology

industry, most recently as President of Biotechnology Associates, a company that provides assistance to American and multinational firms in the development of policies and programs for licensing and technology transfer.

8. Botchan M., Topp W., and Sambrook J. 1978. Studies on simian virus 40 excision from cellular chromosomes. *Cold Spring Harbor Symp. Quant. Biol.* **43:** 709–719 (volume 2).

Cell, Vol. 9, 269–287, October 1976, Copyright © 1976 by MIT

The Arrangement of Simian Virus 40 Sequences in the DNA of Transformed Cells

Michael Botchan, William Topp,
and Joe Sambrook
Cold Spring Harbor Laboratory
P.O. Box 100
Cold Spring Harbor, New York 11724

Summary

High molecular weight DNA, isolated from eleven cloned lines of rat cells independently transformed by SV40, was cleaved with various restriction endonucleases. The DNA was fractionated by electrophoresis through agarose gels, denatured in situ, transferred directly to sheets of nitrocellulose as described by Southern (1975), and hybridized to SV40 DNA labeled in vitro to high specific activity. The location of viral sequences among the fragments of transformed cell DNA was determined by autoradiography. The DNAs of seven of the cell lines contained viral sequences in fragments of many different sizes. The remaining four cell lines each contain a single insertion of viral DNA at a different chromosomal location. The junctions between viral and cellular sequences map at different places on the viral genome.

Introduction

When rodent cells, which are nonpermissive to SV40 multiplication, are infected with the virus, a fraction of them changes its pattern of growth and assumes a new set of stable properties that resemble those of tumor cells (see Tooze, 1973). This process is called transformation.

In general, transformed cells do not synthesize infectious virus and do not contain detectable quantities of the familiar circular forms of viral DNA. Instead they invariably carry viral DNA sequences (Westphal and Dulbecco, 1968) integrated into their chromosomes (Sambrook et al., 1968). Usually the amount of these sequences in the genomes of transformed cells is both small—more than ten viral genome equivalents per diploid quantity of cell DNA is very rare (Gelb, Kohne, and Martin, 1971), while fewer than three is fairly common—and constant— all subclones derived from a given line of transformed cells contain the same quantity of viral DNA as their parents (Ozanne, Sharp, and Sambrook, 1973; Botchan et al., 1974).

Many lines of SV40-transformed cells contain complete copies of SV40, and from these cells, infectious virus often can be rescued using any one of a number of techniques (Gerber, 1966; Watkins and Dulbecco, 1967; Koprowski, Jensen, and Steplewski, 1967). However, integration of the entire viral genome is not obligatory for transformation be-cause cells can be transformed by fragments obtained by digestion of viral DNA with restriction endonucleases (Abrahams and van der Eb, 1975; Abrahams et al., 1975). Analysis by reassociation kinetics shows that such cells can lack virtually all the sequences of the late genes of SV40 (A. J. van der Eb, J. Arrand, and J. Sambrook, unpublished data). This evidence, together with that obtained from physiological studies of cells transformed by temperature-sensitive mutants of the virus (Brugge and Butel, 1975; Tegtmeyer, 1975; Kimura and Hagaki, 1975; Osborn and Weber, 1975; Martin and Chou, 1975), demonstrates that expression of the early gene of SV40 is necessary for establishing and maintaining the transformed state.

Whether integration has any role in this expression is unknown. It is possible that transformation is consequent upon integration of viral DNA at certain sites or in particular orientations within the cellular genome; at the very least, integration must provide a stable environment for the replication of viral DNA and for the expression of its necessary functions.

In this paper, we report on the organization of the viral DNA sequences that are integrated into the genomes of an isogenic set of eleven lines of rat cells independently transformed by SV40. The experiments follow the general design used by Botchan and McKenna (1973) in their original study of the SV40 DNA sequences in two lines of transformed mouse cells. They differ, however, in certain particulars of procedure. High molecular weight DNA, extracted from transformed cells, is cleaved by a sequence-specific restriction endonuclease. The resulting DNA fragments are fractionated by electrophoresis through an agarose gel, denatured in situ, transferred to a nitrocellulose filter as described by Southern (1975), and hybridized to SV40 DNA that has been labeled in vitro to high specific activity by the so-called "nick translation" reaction (Kelly et al., 1970; P. Berg, personal communication). The distribution of viral sequences among the different sized fragments of transformed cell DNA is then determined by autoradiography. The important feature of this method of hybridization is that the labeled viral DNA used as probe is present in vast excess and provides the driving force for the reaction. By this technique, it is easily possible to detect bands that contain as little as 10^{-13} g of SV40 DNA. By analyzing 5 μg of cell DNA, it is possible to recognize as few as 0.02 copies of SV40 DNA per diploid equivalent of transformed cell DNA.

We have used a variety of restriction endonucleases that can be classified into three groups for the purposes of these experiments (see Figure 1). First, there are nucleases such as Bai1 and Smal

ROBERT TJIAN

Tjian R. 1978. **The binding site on SV40 DNA for a T antigen-related protein.** *Cell* **13**: 165–179. (Reprinted, with permission, from Elsevier ©1978.)

N 1975, AS I CONTEMPLATED A POSTDOCTORAL position while finishing my graduate studies in Rich Losick's laboratory,[1] I had the good fortune to talk to Jim Watson at a Harvard Junior Fellows' Tuesday lunch. He convinced me that my interest in gene regulatory proteins would fit in well with work at Cold Spring Harbor Laboratory. Jim also saw that I had a natural affinity for protein biochemistry and so he challenged me to find out how SV40 T antigen might work to regulate DNA replication and transcription.

I arrived at Cold Spring Harbor in the spring of 1976 just after getting married to Claudia in Los Angeles and immediately began growing CV-1 cells and infecting them with SV40. In the first few weeks, I managed to contaminate more cells with *Bacillus subtilis* than to infect them with SV40. But eventually with the help of Terri Grodzicker[2] and later Yasha Gluzman,[2] I shed my "spore" contaminated clothes and began to produce sufficient SV40-infected cells to attempt extracting proteins and SV40 chromatin, following methods described in Bill Sugden's thesis.[3] Like most young naive scientists, I was not fully aware of either the technical challenges or conceptual hurdles facing me. The project seemed to be simple enough: Purify SV40 large T antigen and develop assays to see how the protein works to initiate DNA synthesis and control early/late SV40 transcription, functions anticipated from the phenotype of temperature-sensitive mutants of T antigen.[4]

During the summer of 1976, a number of laboratories presented studies using nonspecific filter-binding assays or low-resolution electron microscopy suggesting

BOB TJIAN is currently Howard Hughes Investigator and Professor of Biochemistry and Molecular Biology in the Division of Molecular and Cellular Biology, Berkeley. Born in Hong Kong, he attended high school in New Jersey, entered Berkeley in 1967, and obtained his B.A. in Biochemistry in 1971. After receiving his Ph.D. in biochemistry and molecular biology from Harvard in 1976, he worked as a postdoctoral researcher at Cold Spring Harbor until 1979. He then returned to Berkeley as an assistant professor in the Department of Molecular and Cell Biology. His major research interest for the past 30 years has been transcription factors—most recently, those that govern whether embryonic stem cells maintain pluripotency or differentiate. jmlim@berkeley.edu

Bob Tjian, 1977. *(Courtesy of Cold Spring Harbor Laboratory Archives.)*

that T antigen might function as some sort of DNA-binding protein. However, the evidence was weak and by the end of the summer's meetings, I had became convinced that T antigen was likely <u>NOT</u> a sequence-specific DNA-binding factor, in large part because I had no luck demonstrating any sequence-specific DNA-binding activity of T antigen isolated from SV80-transformed cells, in those days, a commonly used starting material for extraction of T antigen. To get past these initial failures, I started again and set about systematically building a set of critical reagents and developing new assays. A first important step forward was to generate antibodies specific for T antigen, which I managed with the help of Georg Fey.[5] The next big break was a serendipitous finding that a rather exotic adenovirus-SV40 hybrid virus D2, which had recently been isolated and characterized in James Laboratory, greatly overproduced a hybrid protein closely related to SV40 T antigen.[5] The last and most difficult step was the development of various in vitro biochemical and in vivo functional assays for tracking T antigen through purification and characterizing its properties.

Initially, I relied on a rapid immunoassay to detect D2 protein during column chromatography. To overcome the more challenging problem of a functional assay, I used the rather cumbersome but effective single-cell microinjection system developed and taught to me by Adolf Graessmann.[6] By injecting purified protein into cultured cells and measuring its ability to stimulate cellular DNA replication, I found that the D2 protein was indeed active and functional (unlike the "T antigen" isolated from SV80 cells, which turned out to be mutant and non-DNA binding). The third and most revealing assay was a highly specific DNase protection followed by pyrimidine tract analysis. In this assay, purified D2 protein protects binding sites on SV40 DNA from digestion with DNase. The pattern of pyrimidine residues in the protected DNA identifies the location of the binding sites in the SV40 genome. Mike Botchan[7] taught me nick translation for labeling DNA and Rich Roberts helped with the pyrimidine tract analysis.[8] Thus, I established that SV40 T antigen indeed behaves as a eukaryotic sequence-specific DNA-binding factor that recognizes multiple tandemly arranged sites that overlap the

SV40 origin of DNA replication and the early transcriptional promoter. These studies provided the first definite evidence for the existence and function of a eukaryotic sequence-specific DNA-binding regulatory factor. The results of these studies were compiled into the paper chosen for this volume. A year later, we found that T antigen is also an ATPase (and not a kinase, as had been reported on the heels of the discovery of SRC kinase[9]).

In 1979, after moving to Berkeley, Don Rio[10] and I developed a regulated in vitro transcription system and showed that T antigen selectively repressed SV40 early transcription in vitro by binding to these tandem sites. These experiments not only told us a lot about SV40, but, more importantly, also led directly to a series of key studies aimed at dissecting the transcriptional apparatus of metazoan organisms. Success with SV40 T antigen generated the impetus for a career studying a variety of eukaryotic (human and *Drosophila*) DNA-binding proteins such as the human factor SP1 and a slew of other gene-specific and basal transcription factors.

Starting with the relatively straightforward discovery of how SV40 T antigen works, we have managed to build a rich body of evidence that today includes not only the essential and large family of sequence-specific DNA-binding transcriptional activators/repressors, but also the core promoter initiation complexes (PIC), multisubunit coactivators, cell-type-specific components of the PIC, and a cadre of chromatin-modifying/remodeling cofactors. This massive and elaborate regulatory apparatus must be assembled correctly at each promoter in order to orchestrate the spatial and temporal patterns of gene transcription that define metazoan organisms. In addition, our understanding of the relationship between transcription factors, such as AP1/Jun/Fos, NF-κB, Myc/Max, SREBP1-2, and PPARγ, etc., has significantly enriched our knowledge of how these critical regulatory factors play key roles in diseases ranging from cancer to diabetes, hypercholesterolemia, and inflammation.

Notes and References

1. Rick Losick is best known for his work on spore formation in the soil bacterium *Bacillus subtilis*. The process, which involves asymmetric division of the parental bacterium, is governed by a cascade of five developmental proteins that appear in a specific order over the course of spore development and turn on different sets of genes. Bob Tjian was involved in the early stages of this work, and his thesis project concerned the characterization of factors that bind to *B. subtilis* RNA polymerase and modify its action.

2. In 1976, Terri Grodzicker was a staff member in James Laboratory, who, like Bob Tjian, had recently switched her research interests from prokaryotic to eukaryotic

systems. A year or so later, Yasha Gluzman arrived in James Laboratory from Israel, already an expert on the biology of SV40.

3. Bill Sugden was one of the first occupants of James Laboratory after renovation was completed in 1969. His work over the next four years on RNA polymerase in human cells formed the basis of a Ph.D. awarded by Columbia University in 1973. In his Director's Report of that year, Jim Watson noted that "Bill's work represents the first research done at the Lab that had led to the Ph.D., an event we hope will happen many more time in the future."

4. Cowan K., Tegtmeyer P., and Anthony D.D. 1973. Relationship of replication and transcription of simian virus 40 DNA. *Proc. Natl. Acad. Sci.* **70:** 1927–1930.

5. Georg Fey, a first-class protein chemist, worked at Cold Spring Harbor from 1975 to 1977, mostly in collaboration with a group who were isolating and characterizing a series of novel, naturally occurring adenovirus-SV40 hybrid viruses. In infected cells, one of these hybrids, Ad2$^+$D2, produced very large quantities of a protein with many of the properties of SV40 T antigen. Hassell J.A., Lukanidin E., Fey G., and Sambrook J. 1978. The structure and expression of two defective adenovirus 2/simian virus 40 hybrids. *J. Mol. Biol.* **120:** 209–247.

6. Adolf Graessmann came in 1978 to Cold Spring Harbor with his wife Monika for a six-month sabbatical. Both of the Graessmann's were experts at introducing proteins and nucleic acids into single cells in culture by microinjection. The techniques that they brought to Cold Spring Harbor became a cornerstone of many aspects of the Laboratory's work for the next few years.

7. In nick translation, DNA is copied in an in vitro reaction by an *Escherichia coli* enzyme, DNA polymerase I, in the presence of radioactive nucleotides. In this way, highly radioactive "DNA probes" containing many labeled nucleotides can be produced for subsequent use in nucleic acid hybridization reactions. The biochemical reactions involved in nick translation were defined in Arthur Kornberg's laboratory in Stanford University in the 1960s. By the mid 1970s, nick translation had been adopted by other laboratories at Stanford (notably Paul Berg's) to label nucleic acids in vitro. The reaction was brought to Cold Spring Harbor in 1975 by Tom Maniatis who, in turn, taught Mike Botchan. The full details of the reaction were finally published by the Stanford group in 1977. Rigby P.W., Dieckmann M., Rhodes C., and Berg P. 1977. Labeling deoxyribonucleic acid to high specific activity in vitro by nick translation with DNA polymerase I. *J. Mol. Biol.* **113:** 237–251.

8. Rich Roberts was the best nucleic acid chemist on the staff of the Laboratory in the late 1970s. By that time, pyrimidine tract analysis of DNA had been made almost obsolete by more modern sequencing methods. However, the venerable technique was ideal for Bob Tjian's purposes since it generated DNA fingerprints that unambiguously identified the binding sites for T antigen on SV40 DNA.

9. Tjian R. and Robbins A. 1979. Enzymatic activities associated with a purified simian virus 40 T antigen-related protein. *Proc. Natl. Acad. Sci.* **76:** 610–614.

10. Rio D., Robbins A., Myers R., and Tjian R. 1980. Regulation of simian virus 40 early transcription in vitro by a purified tumor antigen. *Proc. Natl. Acad. Sci.* **77:** 5706–5710.

Cell, Vol. 13, 165–179, January 1978, Copyright © 1978 by MIT

The Binding Site on SV40 DNA for a T Antigen-Related Protein

Robert Tjian
Cold Spring Harbor Laboratory
Cold Spring Harbor, New York 11724

Summary

A protein closely related to SV40 T antigen was purified in a biologically active form from cells infected with the defective adenovirus-SV40 hybrid, Ad2⁺D2. This 107,000 dalton hybrid protein binds and protects a specific portion of SV40 DNA from digestion by pancreatic DNAase I. Hybridization, endonuclease cleavage and pyrimidine tract analysis of the protected fragments reveal that the D2 hybrid protein binds in a sequential manner to tandem recognition sites which lie within a sequence of 120 nucleotides at position 67 near the origin of SV40 replication.

Introduction

The tumor antigen of SV40, a protein of approximately 100,000 daltons found in cells transformed by SV40, reacts specifically with sera raised in animals bearing tumors induced by the virus (Black et al., 1963; Tooze, 1973; Tegtmeyer, 1975). Although genetic studies indicate that T antigen is coded, at least in part, by the A gene of SV40 (Tegtmeyer et al., 1975), DNA sequences which encode the protein have not been precisely mapped. T antigen is involved in a number of functions which include the ability to trigger the onset of cellular DNA synthesis (Graessmann and Graessmann, 1976); to initiate viral replication (Tegtmeyer, 1972); to allow the expression of late SV40 genes (Cowan, Tegtmeyer and Anthony, 1973); to regulate the transcription of its own gene (Tegtmeyer, 1975; Reed, Stark and Alwine, 1976); to provide helper function for the growth of adenovirus type 2 (Ad2) in monkey cells (Rabson et al., 1964; Kimura, 1974; Grodzicker, Lewis and Anderson, 1976); and to initiate abortive as well as stable transformation and perhaps to maintain the transformed phenotype in a variety of cell types (Kimura and Dulbecco, 1972; Tegtmeyer, 1975; Brugge and Butel, 1975; Martin and Chou, 1975; Osborn and Weber, 1975; Steinberg et al., 1978).

How this early gene product of SV40 regulates these diverse physiological events is unknown, but T antigen is known to bind tightly to many kinds of double-stranded DNA (Carrol, Hager and Dulbecco, 1974) and specifically to certain regions of the SV40 genome (Reed et al., 1975; Jessel et al., 1976). Because the regulation of both viral DNA synthesis and transcription could be a direct result of the binding of T antigen to regulatory sequences

on the SV40 genome, I decided to examine further the interaction between T antigen and DNA.

T antigen has been isolated from many sources and extensively characterized (Villano and Defendi, 1973; Livingston, Henderson and Hudson, 1974; Tegtmeyer, Rundell and Collins, 1977), but it has always been difficult to obtain large quantities of pure protein which retained biological activity. To circumvent some of these technical difficulties, a protein closely related to SV40 T antigen was isolated from cells infected with an adenovirus-SV40 hybrid, Ad2⁺D2 (J. Hassell, Y. E. Lukanidin, G. Fey and J. Sambrook, manuscript submitted). The genome of this virus lacks the sequences which map between positions 76 and 96 on the conventional adenovirus map (Figure 1). Inserted in their place is a stretch of DNA encompassing all of the SV40 genome except for the region which maps between positions 54 and 63. The cytoplasm of cells infected with Ad2⁺D2 contains a transcript which starts near position 75 in the adenovirus genome and continues into the SV40 A gene terminating at position 17 within the SV40 sequences. This hybrid transcript codes for a 107,000 dalton protein which contains approximately 10,000 daltons of an unidentified Ad2 protein at its amino terminus and approximately 100,000 daltons of the SV40 A gene protein at its carboxy terminus. The 107,000 dalton protein is specifically immunoprecipitated by antisera from hamsters bearing tumors induced by SV40 (J. Hassell, Y. E. Lukanidin, G. Fey and J. Sambrook, manuscript submitted) and shares extensive peptide homology with SV40 T antigen as judged by analysis of partial proteolytic digestion of both proteins (R. Tjian, unpublished results). Moreover, it is capable of carrying out at least some of the same functions expressed by authentic SV40 T antigen. Thus after injection into cells, the purified 107,000 dalton protein induces the synthesis of cellular DNA and provides helper function for the growth of Ad2 in monkey cells. In addition, cells injected with DNA fragments containing the SV40 sequences from Ad2⁺D2 complement temperature-sensitive A gene mutants as well as deletion mutants of SV40 which lack a functional T antigen (R. Tjian, G. Fey and A. Graessmann, manuscript in preparation). Finally, cells infected with DNA from Ad2⁺D2 complement the growth of temperature-sensitive mutants in the A gene of SV40 (J. Sambrook and J. Hassell, personal communication). Although the 107,000 dalton protein from Ad2⁺D2 shares many of the properties attributed to the SV40 T antigen, to avoid confusion, it will be referred to as the D2 hybrid protein.

The main advantage of using Ad2⁺D2 stems from the finding that cells infected with this virus produce 10–50 times more protein related to the A

JOSEPH F. SAMBROOK

Gluzman Y. 1981. **SV40-transformed simian cells support the replication of early SV40 mutants.** *Cell* **23:** 175–182. (Reprinted, with permission, from Elsevier ©1977.)

YAKOV (YASHA) GLUZMAN WAS 49 YEARS OLD at the time of his death. This paper—his finest—had been published 15 years earlier, when Yasha was a postdoctoral fellow in James Laboratory at Cold Spring Harbor. The idea fueling the work is even older, stemming from Yasha's graduate student days in Ernest Winocour's laboratory at the Weizman Institute in Israel. At that time, the genetics of SV40 was still in a fairly primitive state: The number of viral genes was unknown, the viral genome had not been sequenced, and mutants with useful properties were few in number. Yasha's Ph.D. project was to create lines of cultured monkey cells that supported growth of SV40 and expressed a fully functional SV40 T antigen, the key protein involved in replication of viral DNA and malignant transformation. These cells could then be used in a variety of ways: For example, they could be used as permissive hosts to select and isolate mutants of SV40 with defects in T antigen or to propagate any transfected DNA that carried an SV40 origin of DNA replication.

In his first attempt at the experiment in Israel, Yasha infected simian cells with SV40 that had been inactivated by irradiation with ultraviolet light. He was able to isolate several transformants that seemed to express T antigen and to support growth of a superinfecting mutant virus. But there was a problem. The virus that emerged from the cells was not mutant but wild type generated by recombination between the superinfecting mutant and the resident integrated copy of the defective viral genome.[1]

Yasha was both persistent and stubborn, and by the time he arrived in Cold Spring Harbor was keen to try the experiment again. By then (1978), the ill-advised moratorium on molecular cloning of tumor virus genomes and cellular oncogenes had been lifted and the DNA sequence of SV40 was known, and Yasha was able to try a much more sophisticated and watertight version of his experiment.

JOE SAMBROOK (*See biographical footnote on p. 37.*)

Yasha Gluzman, 1983. *(Courtesy of Cold Spring Harbor Laboratory Archives.)*

The restriction enzyme *Bgl*I cleaves SV40 DNA just once in the region of the viral genome known as the origin, where controlling elements are clustered and replication of viral DNA is initiated. Yasha and his colleagues in James Laboratory used this information to isolate a set of deletion mutants lacking the *Bgl*I site and surrounding DNA sequences essential for origin function. These mutants could still synthesize active T antigen but were incapable of growth in mammalian cells.[2,3] Yasha used one of these origin-minus mutants that lacked just 6 base pairs of SV40 DNA to create lines of cultured simian CV-1 cells (COS cells)[4] whose genomes now contained integrated copies of the viral DNA. These cells did everything that Yasha had hoped: They expressed wild-type T antigen in sufficient quantity to support replication of SV40 mutants whose own T-antigen gene was defective and to drive replication of sequences of foreign DNAs attached to the SV40 origin of replication.

COS cells had another important and timely property. By the beginning of the 1980s, it had become abundantly clear that manufacturing mammalian proteins in native form in bacteria—especially large proteins—was a far more difficult task than many had anticipated. As a consequence, interest quickened in developing mammalian vectors and appropriate lines of host cells. COS cells turned out to have far more uses than even Yasha could have foreseen: They are efficient and malleable hosts for transient expression of a very wide range of transfected genes and for amplified expression of mammalian coding sequences attached to plasmids carrying an SV40 origin of replication.

Yasha left Cold Spring Harbor in 1987, and for the last nine years of his life, he worked on a wide variety of topics—from the mode of action of antibiotics to the development of antiviral drugs, to the genetic analysis of herpesviruses. But his most enduring legacy, and his best piece of science, was the development and characterization of COS cells. Long after his death, COS cells continue to be a mainstay of mammalian gene expression and analysis. As of November 2005, PubMed contained close to 18,000 references to publications that have used these cells.

Yasha Gluzman enjoyed his science hugely. Solid, muscular, and with the build of a budding sumo wrestler, his ebullience and irrepressible good humor were welded to an eager analytical mind and an ability to solve difficult scientific problems with strings of simple explicit experiments. An emotional and sensual man, he fitted Macauley's description of "reason penetrated and made red hot by passion."[5]

Notes and References

1. Gluzman Y., Davison J., Oren M., and Winocour E. 1977. Properties of permissive monkey cells transformed by UV-irradiated simian virus 40. *J. Virol.* **22:** 256–266.

2. Gluzman Y., Sambrook J.F., and Frisque R.J. 1980. Expression of early genes of origin-defective mutants of simian virus 40. *Proc. Natl. Acad. Sci.* **77:** 3898–3902.

3. Gluzman Y., Frisque R.J., and Sambrook J.F. 1980. Origin-defective mutants of SV40. *Cold Spring Harbor Symp. Quant. Biol.* **44:** 293–300.

4. COS is an acronym of CV-1, Origin-minus, SV40.

5. Thomas Babington Macaulay, The Edinburgh Review, July 1835.

Cell, Vol. 23, 175–182, January 1981, Copyright © 1981 by MIT

SV40-Transformed Simian Cells Support the Replication of Early SV40 Mutants

Yakov Gluzman
Cold Spring Harbor Laboratory
P.O. Box 100
Cold Spring Harbor, New York 11724

Summary

CV-1, an established line of simian cells permissive for lytic growth of SV40, were transformed by an origin-defective mutant of SV40 which codes for wild-type T antigen. Three transformed lines (COS-1, -3, -7) were established and found to contain T antigen; retain complete permissiveness for lytic growth of SV40; support the replication of tsA209 virus at 40°C; and support the replication of pure populations of SV40 mutants with deletions in the early region. One of the lines (COS-1) contains a single integrated copy of the complete early region of SV40 DNA. These cells are possible hosts for the propagation of pure populations of recombinant SV40 viruses.

Introduction

Some of the virus-specific proteins expressed in cells transformed by DNA tumor viruses are responsible for maintenance of the transformed phenotype (Tooze, 1980). Furthermore, viral proteins synthesized in transformed permissive cells can be utilized to complement the growth of superinfecting mutant viruses. For example, permissive cell lines transformed by adenoviruses (Shiroki and Shimojo, 1971; Graham et al., 1977; Grodzicker and Klessig, 1980) and herpes simplex virus (Kimura et al., 1974; Macnab and Timbury, 1976) have been isolated. Using such cells as permissive hosts, host-range mutants of Ad5 (Harrison et al., 1977), deletion mutants of Ad5 (Jones and Shenk, 1979), ts mutants of Ad12 (Shiroki et al., 1976) and ts mutants of herpes simplex virus type 2 (Kimura et al., 1974; Macnab and Timbury, 1976) were successfully propagated because of complementation by viral proteins specified by the integrated viral sequences. However, attempts to develop a similar system for papovaviruses have not been as successful. For example, simian cells transformed by SV40 are either resistant to superinfection (Shiroko and Shimojo, 1971; Gluzman, unpublished data) or are unable to complement the growth of defective viruses (Gluzman et al., 1977a). More encouraging results have been obtained with polyoma virus; host-range mutants have been isolated that grow better on polyoma-transformed mouse cells than on their untransformed counterparts (Benjamin, 1970). However, the enhanced growth of these mutants was not due to direct complementation by viral products expressed in the polyoma-transformed mouse cells (Goldman and Benjamin, 1975).

This paper describes the development and properties of a new line of simian cells, obtained by transformation of CV-1 cells with an origin-defective mutant of SV40 (Gluzman et al., 1980a, 1980b). These cells are permissive to infection with wild-type SV40. They support replication of the DNA of SV40 tsA mutants at nonpermissive temperature, as well as the growth of pure populations of defective viruses with deletions in the early region.

Results

Transformation of CV-1 Cells by SV40 Origin-Defective Mutant DNA

Transformation was carried out using SV40 DNA cloned in plasmids and propagated in E. coli. The plasmid DNAs contain either the origin-defective mutant 6-1 or wild-type. The wild-type SV40 recombinant (wild-type plasmid) DNA consists of a complete SV40 genome inserted into the Bam HI site of the modified vector pMK16 #1. The mutant 6-1 DNA was derived from this wild-type plasmid DNA by deleting six nucleotides at the origin of viral DNA replication (Gluzman et al., 1980a, 1980b). Freshly seeded monolayers of CV-1 or TC7 cells (~5 × 10⁶ cells/10 cm plate) were transfected with 6-1 or wild-type plasmid DNA using either the DEAE-dextran or calcium technique and the cultures treated as described in Experimental Procedures. Both CV-1 and TC7 cells transfected with wild-type SV40 plasmid DNA eventually gave rise to infectious virus, and most of the cultures lysed after 6 weeks; the DNA of the released infectious viruses contained either small insertions of pMK16 #1 DNA, originating from the vector–SV40 Bam HI junction, or small deletions of SV40 sequences around the Bam HI site (Y. Gluzman, manuscript in preparation). The morphology of the cells on the few plates of CV-1 or TC7 cells that survived transfection with the wild-type plasmid remained indistinguishable from mock-infected cultures.

No infectious virus was detected in cells transfected with 6-1 DNA. Transformed colonies were first observed at the end of the fourth week and after one and a half months their number reached 10–20 colonies/plate. Transformed colonies emerged only in cultures of CV-1 cells transfected with 6-1 DNA using the Ca⁺⁺ precipitation technique. Transformed foci, composed of densely growing but flat cells, were isolated and transferred into 2.5 cm plates. The mass cultures, derived from individual colonies, contain varying proportions of T antigen-positive cells. Attempts to subclone T antigen-positive cells from early passages were unsuccessful, and all isolated single colonies were T antigen-negative. However, after passaging the mixed cultures of transformed and normal cells for 1–2 months, more than 98% of the cells were T antigen-positive. Three of these uncloned cell lines were chosen for further work and were designated

Thummel C., Tjian R., and Grodzicker T. 1981. **Expression of SV40 T antigen under control of adenovirus promoters.** *Cell* **23:** 825–836. (Reprinted, with permission, from Elsevier ©1981.)

J UST AS BACTERIOPHAGES PROVIDED A SIMPLE MODEL system for defining the basic principles of molecular biology, DNA tumor viruses, a major focus of study in the late 1970s and early 1980s, provided our first insights into the mechanisms of eukaryotic transcription and gene regulation. This work was in full flourish when I met my coauthors on this paper, Terri Grodzicker (see Grodzicker, p. 99) and Robert Tjian (see Tjian, p. 119) (better known as Tij), during the summer of 1977. I had been fortunate enough to get a position at Cold Spring Harbor Laboratory that summer, gaining research experience before entering graduate school at the University of California, Berkeley. Tij was busy at the time studying the early region of SV40, which encoded large T antigen, one of the first eukaryotic transcription factors to be characterized. Although SV40 DNA could be readily purified, providing a substrate for DNA-binding studies, a source of native T antigen was more difficult to obtain. Tij's work at Cold Spring Harbor exploited a naturally occurring adenovirus–SV40 hybrid virus, Ad2+D2, that overexpressed a fusion protein consisting of 10 kD of an adenoviral structural protein replacing the amino terminus of T antigen.[1] It was a straightforward task to prepare large amounts of the D2 protein for biochemical analysis because this protein was so abundant and Ad2+D2 could be grown in suspension cultures of HeLa cells. Purification of native T antigen was, by comparison, both expensive and laborious, requiring at least 100 plates of SV40-infected monkey cells. A number of key studies were published that used D2 protein as a surrogate for T antigen, demonstrating specific binding to the SV40 early promoter[2] and negative regulation of the SV40 early promoter in vitro.[3] Working with an unnatural fusion protein, however, was a constant source of concern and provided the basis for my gradu-

---

CARL S. THUMMEL holds a Bachelor's Degree from Colgate University and a Ph.D. from the University of California, Berkeley, where his supervisor was Bob Tjian. After postdoctoral work at Stanford with David Hogness, Carl joined the Department of Human Genetics, University of Utah in 1987, where he is now a Professor. carl.thummel@genetics.utah.edu

Carl Thummel. *(Courtesy of Carl Thummel.)*

ate research when I joined Tij's lab as a graduate student in 1979. We sought to create our own adenovirus–SV40 hybrid viruses with two goals in mind: (1) to provide a source for purifying native T antigen for biochemical studies and (2) a more general goal, to develop a new vector system for overexpressing eukaryotic proteins. We achieved both of these objectives in the paper published below. Future studies took advantage of the T antigen purified from the recombinant viruses we created, both confirming earlier studies with D2 protein and providing new insights into T-antigen function.[4,5] In addition, our work set the stage for the development of adenovirus as a eukaryotic vector system. It is gratifying to see that adenovirus is used routinely today for gene expression in mammalian cells, as well as for human gene therapy trials, a potentially revolutionary clinical application that we did not fully appreciate at the time we were doing our work.

Cold Spring Harbor's influence over our science persisted after our move to the West Coast. Tij arrived at Berkeley in January of 1979 and moved into an empty lab with two graduate students, Rick Myers[6] and myself. Rather than gathering lab supplies and equipment in the traditional fashion, Tij's idea for setting up the lab was to do a D2 protein purification, having arrived with a batch of infected cells all ready to go. This certainly accomplished the goal in a speedy manner and also introduced us to the ritual of round-the-clock research that was to characterize the Tjian lab for years to come. I was particularly fortunate to have a second advisor contribute to my graduate training, our collaborator from Cold Spring Harbor, Terri Grodzicker. Terri was an expert in adenovirus biology and genetics and was responsible for developing our approaches to create designer adenovirus–SV40 hybrids. Along with Tij, she provided outstanding guidance on how to conduct scientific research and how to present the results, both orally and in writing. In retrospect, I cannot imagine a better graduate education, although it could occasionally be a challenge to meet the high standards of both Tij and Terri. For me, this paper epitomizes the exciting research that was conducted during the first 5 years of the Tjian lab at Berkeley and the heady atmosphere of modern molecular genetics—that whole genomes could now be manipulated in a precise way and for a specific purpose.

Notes and References

1. Hassell J.A., Lukanidin E., Fey G., and Sambrook J. 1978. The structure and expres-

sion of two defective adenovirus 2/simian virus 40 hybrids. *J. Mol. Biol.* **120:** 209–247.

2. Tjian R. 1978. The binding site on SV40 DNA for a T antigen-related protein. *Cell* **13:** 165–179.

3. Rio D., Robbins A., Myers R., and Tjian R. 1980. Regulation of simian virus 40 early transcription in vitro by a purified tumor antigen. *Proc. Natl. Acad. Sci.* **77:** 5706–5710.

4. Myers R.M., Rio D.C., Robbins A.K., and Tjian R. 1981. SV40 gene expression is modulated by the cooperative binding of T antigen to DNA. *Cell* **25:** 373–384.

5. Jones K.A. and Tjian R. 1984. Essential contact residues within SV40 large T antigen binding sites I and II identified by alkylation-interference. *Cell* **36:** 155–162.

6. Rick Myers is now Professor and Chair of the Department of Genetics at Stanford University.

Cell, Vol. 23, 825–836, March 1981, Copyright © 1981 by MIT

Expression of SV40 T Antigen under Control of Adenovirus Promoters

Carl Thummel and Robert Tjian
Department of Biochemistry
University of California
Berkeley, California 94720
Terri Grodzicker
Cold Spring Harbor Laboratory
Cold Spring Harbor, New York 11724

Summary

We have obtained novel adenovirus–SV40 recombinant viruses that express wild-type SV40 large T
and small t antigens under the control of different
adenovirus promoters. Hybrids were constructed in
vitro with SV40 DNA that contains the entire early
coding region but lacks the transcriptional promoter. Recombinants were isolated by a strong
biological selection for viruses that express SV40 T
antigen. Analysis of several recombinant genomes
indicates that they contain the SV40 A gene inserted in a variety of positions and orientations in
the adenoviral genome. Moreover, the set of hybrid
transcripts reveals an unexpected variety of splicing patterns. Some hybrid mRNAs transcribed from
the adenovirus late promoter appear to contain the
adenovirus tripartite leader sequence. Other hybrid
mRNAs were transcribed from adenovirus early promoters. All recombinant mRNAs contain intact SV40
early sequences that have normal splice patterns
and produce wild-type T antigens. Biochemical
characterization of SV40 T antigens overproduced
by the hybrid viruses indicates that they are structurally indistinguishable from wild-type SV40 large
T antigen and are functionally equivalent to the D2
protein.

Introduction

In order to facilitate the isolation and characterization
of specific eucaryotic regulatory proteins, we have
developed a human adenovirus vector system designed to express foreign genes under viral transcriptional control. As a prototype, we chose to insert the
SV40 A gene into adenovirus in order to create a
hybrid virus capable of synthesizing high levels of
wild-type SV40 large T antigen. The A gene product
of SV40 is a 96,000 dalton nuclear phosphoprotein
that is normally synthesized in SV40-transformed cells
and early during lytic infection of monkey cells (Tooze,
1980). This early gene product is involved in the
initiation of viral DNA replication (Tegtmeyer, 1972),
regulation of viral transcription (Cowan et al., 1973;
Reed et al., 1976; Rio et al., 1980) and induction of
transformation (Tegtmeyer, 1975). Furthermore, T antigen provides a helper function that allows adenovirus
to be propagated efficiently on otherwise nonpermissive monkey cells (Rabson et al., 1964; Levine et al.,

1973; Grodzicker et al., 1976). We have exploited
this helper activity in our isolation of adenovirus–SV40
recombinants, by applying it as a direct biological
selection for hybrid viruses that express T antigen in
monkey cells.

In the past, it has been difficult to study and purify
SV40 large T antigen because of its low level of
synthesis in SV40-transformed or lytically infected
cells. Instead, we have characterized a related protein
purified from cells infected with a naturally occurring
adenovirus–SV40 hybrid virus, Ad2$^+$D2 (Hassel et al.,
1978; Tjian, 1978). This virus codes for a 107,000
dalton fusion protein that contains 10,000 daltons of
an adenovirus structural protein replacing the amino-
terminal portion of T antigen. Because the D2 protein
is a hybrid polypeptide, it is important to determine
the biochemical properties of authentic T antigen and
compare them with those of the D2 protein.

Here we report the in vitro construction of adenovirus–SV40 hybrid viruses that express the authentic
SV40 early gene products, large T and small t antigens, under adenovirus transcriptional control. In recent years a wealth of genetic, structural and biochemical information has been gathered concerning
the human adenoviruses (Tooze, 1980). These linear
(35,000 bp) double-stranded DNA tumor viruses undergo a temporally regulated lytic cycle in permissive
human cells. Adenoviruses contain a strong transcriptional promoter, which acts after the onset of viral
DNA replication to efficiently transcribe a large block
of late viral genes (Backenheimer and Darnell, 1975;
Chow et al., 1977). These primary late transcripts are
subsequently processed by RNA splicing to produce
a collection of mRNAs, each containing a common
tripartite leader sequence (Berget et al., 1977; Broker
et al., 1977). Our knowledge of adenovirus transcription and RNA splicing provides a solid foundation for
testing the effects of foreign gene sequences on these
processes. In addition, adenovirus selectively inhibits
host cell protein synthesis late in infection, thereby
simplifying the detection and purification of viral proteins expressed under the control of the late promoter
(Ginsberg et al., 1967; Russell and Skehel, 1972).
Finally, adenovirus-infected human cells can be grown
in spinner culture, which greatly facilitates the isolation of infected cells for preparative purposes.

Purification of the SV40 large T antigen produced
by the hybrid virus has enabled us to study the biochemical properties of this multifunctional regulatory
protein. In addition, analysis of the adenovirus–SV40
hybrid mRNAs has revealed an unexpected flexibility
in the splicing of primary transcripts. The ligation of
foreign eucaryotic genes of interest to the SV40 early
region and their subsequent insertion into the adenovirus vector system, coupled with the selection of T
antigen helper function, should be useful for the cloning and expression of a variety of mammalian genes.

WINSHIP HERR

Herr W. and Clarke J. 1986. **The SV40 enhancer is composed of multiple functional elements that can compensate for one another.** *Cell* **45:** 461–470. (Reprinted, with permission, from Elsevier ©1986.)

T HIS PAPER, STRANGELY ENOUGH, EVOLVED FROM A SEMINAR that for reasons long forgotten, I failed to attend. I had come to Cold Spring Harbor Laboratory in January 1983 as an "independent" postdoc, which meant that I could work on what I wanted but first had to overcome the challenge of finding an empty piece of lab bench. I set up shop in a room used to teach summer courses on the upper level of James Laboratory, sharing the space with Ed Harlow.[1] Before my arrival, I had been a graduate student with Wally Gilbert at Harvard studying murine retroviruses and had just spent a six-month "vacation" postdoc with Fred Sanger at the Medical Research Council Laboratory of Molecular Biology in Cambridge, England, where I learned the "other way" of DNA sequencing.[2] I had decided to come to Cold Spring Harbor because my fellow graduate student Doug Hanahan[3] spoke glowingly of the Lab's scientific environment and because, seeking advice, as at other times in my career, from my undergraduate mentor Harry Noller,[4] I was told that Cold Spring Harbor was the place to go. There was only one detail missing from my independent postdoc idea—I did not know what I wanted to do.

I arrived in James Laboratory as the new kid on the block, knowing how to sequence DNA well but with no concrete idea of what to work on. So I guess it was not too smart to miss a seminar, especially one given by Alex Rich[5] from MIT who spoke about a rather mysterious DNA structure—Z DNA—within the region of the SV40 genome that controlled transcription into RNA. This region contained a recently discovered[6] regulatory element—called the enhancer—that amazingly

WINSHIP HERR received his Ph.D. from Harvard University in 1982 for studies on recombinant retroviruses in leukemogenic mice with Walter Gilbert. After postdoctoral studies with Fred Sanger in Cambridge, England and at Cold Spring Harbor Laboratory, he joined the Cold Spring Harbor Laboratory faculty in 1984. He served as Assistant Director of the Laboratory from 1994 to 2002 and from 1998 to 2004 was the founding Dean of the Watson School of Biological Sciences. In 2004, he moved to the Centre of Integrative Genomics at the University of Lausanne, where he works on regulators of the human cell-proliferation cycle. Winship.Herr@unil.ch

had the ability to increase the volume of transcription of genes from a very great distance. Alex Rich described how two different sites within the SV40 enhancer had the potential to form left-handed helical Z DNA.[7] A possible link between an unconventional DNA conformation and the mysterious transcriptional enhancers was very exciting indeed. At that time, no one understood how gene expression was regulated in mammalian cells; perhaps, Z-DNA sequences were the elusive volume control.

The day after the seminar, Yasha Gluzman,[8] who had worked on SV40 since his days as a graduate student at the Weizmann Institute, proposed a collaboration: I could use my DNA expertise to mutate and disrupt the Z-DNA-forming potential of the two proposed Z-DNA sites and Yasha would assay the effects on SV40 T-antigen[9] expression and viral growth. Yasha offered a carrot: Get the experiment done in time to present at a meeting on enhancers that he was organizing six weeks later at the Laboratory's Banbury Conference Center, with all the big names in attendance, including Alex Rich.

By the time of the Banbury meeting, we had what we referred to as our "pre-pro-result"—mutation of the predicted Z-DNA sites diminished T-antigen expression significantly. Alex Rich was very pleased and gave me one minute at the end of his talk to describe our results.

After the meeting, the work went downhill and uphill in ways that I had no way of predicting. Downhill because I was never able to show that our mutations had in fact affected the ability to form Z DNA, even using a chemical probing technique that I developed specifically for that purpose,[10] and uphill as a result of Yasha's efforts. First, he showed that besides affecting T-antigen expression, the "anti-Z-DNA" mutations blocked growth of SV40 in permissive cells. Second, he discovered that, after some weeks, revertants appeared whose ability to grow had been restored.

I vividly remember Yasha telling me the news. At first I did not appreciate their import, until Yasha looking directly at me said "Revertants are always interesting." With this encouragement, I set out to find out what changes in the SV40 DNA had occurred to restore growth of the revertant viruses. Here, my DNA sequencing skills came in handy. I analyzed 18 revertants and discovered that they all carried tandem duplications of the mutated enhancer region. Although the duplications differed in size, they all overlapped a 15-bp sequence that encompassed the core element of the enhancer. The duplications all restored viral growth and enhancer function. Yasha and I concluded in our letter to *Nature*[11] that "the consistent duplication of the 'core' region in each of the 18 revertants suggests that these sequences are compensating for and apparently acting independently of the mutated Z-DNA sequences."

For me, this viral revertant analysis was an epiphany. I had never before seen a living entity—if only a virus!—respond before my eyes and tell me something about how biology worked. While these studies were in progress, Yasha returned to his interests in T antigen and I was offered a Staff Investigator position at the Lab, which I accepted, deciding to continue working on enhancers.

I hired a recent college graduate, Jennifer Troge (née Clarke), to be my research assistant and we asked what would happen if we mutated the 15-bp sequence at the core of the SV40 enhancer. To our great pleasure, the mutated virus, like those lacking functional Z-DNA sequences, at first did not

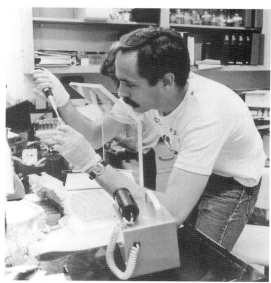

Winship Herr, 1984. *(Courtesy of Cold Spring Harbor Laboratory Archives.)*

grow but eventually generated revertants harboring tandem duplications. This time, however, instead of overlapping the mutated core sequence, the duplications contained one or both of the two "Z-DNA" sites. From this result, we concluded that the SV40 enhancer was composed of three elements, which we called A, B, and C, that could functionally compensate for one another. These results showed that eukaryotic regulatory elements were both modular in structure and surprisingly flexible in design.

When I first submitted the work for publication in *Cell,* the paper was rejected by editor Benjamin Lewin, against the advice of the reviewers. The rejection occasioned my first phone call to an editor to argue my case. Lewin's argument was that the paper published previously in *Nature*[11] with Yasha had upstaged this one. I nervously explained that the earlier paper had told us little about the modularity, redundancy, or plasticity of eukaryotic transcriptional regulatory elements, the essence of this study. After reflection, Lewin graciously agreed to publish the paper.

As I now sit 20 years later in my office at the University of Lausanne gazing at the snowcapped Alps illuminated by a setting winter sun over Lake Geneva, I reflect warmly on how personal interactions during a 22-year stint at Cold Spring Harbor Laboratory greatly influenced my career, especially that fateful winter day in 1983 when Yasha Gluzman proposed a six-week experiment to study Z DNA within the SV40 enhancer.

Notes and References

1. Ed Harlow came to Cold Spring Harbor in 1982, just after he had completed his Ph.D. at the Imperial Cancer Research Fund, London. He left the Laboratory in 1991 to take up positions at Massachusetts General Hospital and Harvard Medical School, where he is now Chair of the Department of Biological Chemistry and Molecular Pharmacology. Ed Harlow's major scientific achievement while at Cold Spring Harbor was his discovery of a physical association between a tumor suppressor protein (Rb) and an adenovirus-encoded oncogenic protein (E1a). For a detailed account of this work, see p. 203.

2. Wally Gilbert and his Harvard colleague Alan Maxam and Fred Sanger at Cambridge independently developed methods to sequence of DNA. The more complex and technically demanding Maxam and Gilbert method used base-specific cleavage chemical cleavage. The Sanger method, much more amenable to automation, used base-specific termination of DNA synthesis in a reaction catalyzed by a DNA polymerase. Both Gilbert and Sanger were awarded Nobel Prizes in Chemistry in 1980 for their work on DNA sequencing.

3. Doug Hanahan worked peripatetically at Cold Spring Harbor between 1978 and 1983, while he was a graduate student at Harvard. He was a full-time member of the Cold Spring Harbor staff between 1983 and 1988, when he joined the Department of Biochemistry and Biophysics at the University of Califormia, San Francisco. At Cold Spring Harbor, Doug generated lines of genetically modified mice that develop pancreatic cancer at high frequency. This work laid the foundation for much of Doug's later career (see p. 193).

4. Harry Noller at the Santa Cruz campus of the University of California has worked for more than 35 years on the structure and function of ribosomes, in particular the structure of ribosomal RNA and its role in protein synthesis.

5. Alex Rich, the Sedgwick Professor of Biophysics at Massachusetts Institute of Technology, is celebrated for solving the three-dimensional structure of transfer RNA as well as for his discovery of Z DNA.

6. Moreau P., Hen R., Wasylyk B., Everett R., Gaub M.P., and Chambon P. 1981. The SV40 72 base repair repeat has a striking effect on gene expression both in SV40 and other chimeric recombinants. *Nucleic Acids Res.* **9:** 6047–6068.

   Banerji J., Rusconi S., and Schaffner W. 1981. Expression of a beta-globin gene is enhanced by remote SV40 DNA sequences. *Cell* **27:** 299–308.

7. Z-DNA is a high-energy form of DNA in which the double helix winds to the left in a zigzag pattern, instead of to the right, as in the more familiar B-DNA.

8. Yasha Gluzman worked at Cold Spring Harbor between 1977 and 1987, first as a postdoctoral fellow and later as a member of staff. For more details, see p. 125.

9. SV40 T antigen is a protein encoded by the "early" region of SV40 DNA. The SV40 enhancer and Z-DNA sequences lie a short distance upstream of the coding sequences for T antigen.

10. Herr W. 1985. Diethyl pyrocarbonate: A chemical probe for secondary structure in negatively supercoiled DNA. *Proc. Natl. Acad. Sci.* **82:** 8009–8013.

11. Herr W. and Gluzman Y. 1985. Duplications of a mutated simian virus 40 enhancer restore its activity. *Nature* **313:** 711–714

Cell, Vol. 45, 461–470, May 9, 1986, Copyright © 1986 by Cell Press

The SV40 Enhancer Is Composed of Multiple Functional Elements That Can Compensate for One Another

Winship Herr and Jennifer Clarke
Cold Spring Harbor Laboratory
Cold Spring Harbor, New York 11724

Summary

We present evidence that the SV40 enhancer consists of three functional units, A, B, and C, each of which can cooperate with the others or with duplicates of itself to enhance transcription. We show that, when element C, containing the core consensus sequence, is inactivated by point mutations, revertants with restored enhancer function contain duplications of either one or both of the elements A and B. To search for additional elements, we isolated revertants of a mutant with the three elements mutated. These revertants do not identify any other elements; instead, enhancer function is effectively restored by "double duplications," in which the first duplication event either partially or entirely recreates one of the three elements A, B, and C and the second duplication then creates two copies of the newly created sequence.

Introduction

Enhancers are promoter elements that can activate transcription over large distances from either upstream or downstream of the transcriptional initiation site (see Gluzman and Shenk, 1983; Gluzman, 1985, for reviews). The mechanism by which enhancers activate transcription is not known, and no nucleotide sequence common to all enhancers has been identified. Nevertheless, a number of short consensus sequences are shared by different sets of enhancers, and enhancers often contain more than one such consensus element. Among the different consensus sequences are the 'core' consensus sequence GTGG$^A/_T$$^A/_T$$^A/_T$G (Laimins et al., 1982; Weiher et al., 1983), and stretches of alternating purines and pyrimidines (Pu/Py) (Nordheim and Rich, 1983).

The enhancer that lies within the early promoter of SV40 was the first for which the unusual positional flexibility of these elements was described (Banerji et al., 1981; Moreau et al., 1981). The enhancer region of the wild-type SV40 strain 776 contains a tandem duplication of 72 bp. The 72 bp element contains both a core consensus sequence, called the SV40 core element (Weiher et al., 1983), and an 8 bp Pu/Py sequence (Figure 1); just upstream of the 72 bp repeat lies a second 8 bp Pu/Py region of different sequence that overlaps another core consensus sequence.

We recently described an enhancer mutant of SV40, *dpm*12, that contains a single 72 bp element and two base substitutions within each of the enhancer Pu/Py segments (Herr and Gluzman, 1985); the combined effect of these four base changes is impaired SV40 enhancer function and SV40 virus growth in the simian cell line CV-1. Analysis of revertants of *dpm*12 (derived by passage of mutant virus stocks in CV-1 cells) showed that tandem duplications of the mutated enhancer region, ranging in size from 45 bp to 135 bp, could restore enhancer function. The new sequences created at the junctions of the different tandem duplications did not exhibit any obvious homology to one another, indicating that these new sequences were probably not responsible for the restoration of enhancer function in the revertants. Instead, the most striking feature of the duplication patterns was the presence of a 15 bp sequence containing the SV40 core element in all of the duplicated regions. These observations led us to suggest that this 15 bp region, which we now call the C element, could compensate for, and function independently of, the mutated Pu/Py sequences.

These experiments left unresolved the relative roles of these various SV40 enhancer sequences in activating transcription. For instance, the 15 bp C element could confer a unique and required function to enhance transcription, while the sequences mutated in *dpm*12 play only an auxiliary role (or roles). Alternatively, the SV40 enhancer could contain multiple independent elements, each capable of enhancing transcription. To discriminate between these and other possibilities, we have made deleterious point mutations within the C element and have determined the enhancer structure of SV40 growth revertants of this mutant, called *dpm*6. These revertants identify two new regions, which we call A and B, each of which spans one of the two separate sequences mutated in *dmp*12. To search for other potential SV40 enhancer elements, we isolated revertants of the triple mutant *dpm*126 in which the three elements A, B, and C are mutated. These revertants do not identify any new elements but indicate that enhancer function can be restored by recreating one of the wild-type sequences changed by the point mutations followed by duplication of the restored sequence.

Results

Point Mutations within the SV40 Enhancer

Figure 1 illustrates the structure of the SV40 early promoter and shows the three sets of double point mutations, 1, 2, and 6, that we have used to debilitate SV40 enhancer function. The point mutations were all introduced into an enhancer lacking one copy of the repeated 72 bp element found in SV40 strain 776. The location of the two 8 bp Pu/Py sequences is indicated in Figure 1 by the hatched boxes marked a and b. In this report, we refer to the 8 bp Pu/Py and core consensus sequences by the lower case letters a, b, and c, and to the three functional enhancer elements by the upper case letters A, B, and C. The pairs of transversion mutations within each of the a and b Pu/Py stretches have been described previously (Herr and Gluzman, 1985). SV40 mutants carrying the different sets of double point mutations are referred to as *dpm* (for double

NEUROSCIENCE

Neuroscience at CSH Laboratory: Meetings, Courses, and Research

Eric Kandel

———

WHAT HAS MADE THE COLD SPRING HARBOR LABORATORY under Jim Watson so magical for so many is its international scope and its youthful, bold, forward-looking perspective. It is, for many people throughout the academic world, a second university: an academic home away from their own academic home. The Lab elicits this sense of belonging because it is at once unpretentious and welcoming socially, as well as interesting scientifically. It distributes its store of scientific knowledge widely through courses, meetings, symposia, and books. With these activities, the Lab makes the recent scientific advances, and the new directions that emerge from these advances, available to scientists throughout the world, irrespective of seniority or age. These activities of the Lab have long reflected the vision of Jim Watson and his unusual and remarkable capacity to point the books, meetings, and courses not toward the past, but toward the future, toward new directions in science.

Just as this is true for molecular biology, so, beginning in the early 1970s, it was true for neurobiology. As someone who has ben-

ERIC KANDEL is University Professor of Physiology and Psychiatry at the Center for Neurobiology and Behavior of the Columbia University College of Physicians & Surgeons and Senior Investigator of the Howard Hughes Medical Institute. He graduated from Harvard College, majoring in history and literature, and received his medical degree from New York University School of Medicine. Before moving to Columbia as founding Director of the Center for Neurobiology and Behavior, he held faculty positions at Harvard and New York Universities. Among many awards, he received the Nobel Prize in Physiology or Medicine in 2000.

efited greatly from my interactions with the Lab and with Jim, let me describe some examples of my experiences with the development of neural science at the CSH Laboratory during the period 1970 to 1990.

Until the genetic code and protein synthesis were solved, Jim Watson's focus at Harvard was on the Central Dogma: How DNA was transcribed into mRNA and how the message was translated into protein. Although most of biology remained in the DNA world for some time afterward, Jim realized that molecular biology was now ready to explore higher-order problems: cellular differentiation, development, and oncogenesis.

Once he assumed the directorship of CSH Laboratory, Jim was encouraged by his friends Seymour Benzer, Sidney Brenner, Cyrus Leventhal, and Gunther Stent—scientists who were transforming themselves from molecular biologists into neurobiologists—to offer summer courses in neurobiology at Cold Spring Harbor for molecular biologists. Many young molecular biologists, Jean-Pierre Changeux, Regis Kelly, Louis Reichardt, and Steve Heineman among them, were now becoming interested in the nervous system. Courses at Cold Spring Harbor could introduce them to the facts and methods of neurobiology. Much as Max Delbrück had retrained himself from physics to molecular biology, so CSH Laboratory might retrain molecular biologists to some day apply molecular techniques to the study of the brain.

In 1975, with funds from the Sloan Foundation in hand, Jim asked Dick Cone at Hopkins, John Nichols and David Hubel at Harvard, and me to help him organize the first set of courses in neural science. We proposed two courses that started that year: one, a lecture course that provided a broad overview of neurobiology, and the other, a laboratory-based course using the simple nervous system of *Aplysia* and its large identified cells that provided a focused introduction to cellular electrophysiological methods. The basic principles course was organized by Regis Kelly, a molecular biologist who had started doing postdoctoral training at the Harvard Department of Neurobiology. He had the help of David Hubel, John Nichols, Eric Frank, Jim Hudspeth, and David Van Essen, all from the Harvard Medical School. In addition, Steve Kuffler, Max Cowan, Zack Hall, Jack McMann, Roger Sperry, David Hubel, and I came to give one or more lectures. The experimental course was led by Philip Ascher and Jac Sue Kehoe from the Ecole Normale in Paris.

In the summer of 1972, a course on the behavioral genetics of *Caenorhabditis elegans* was added, and in 1973, a course on the neurobiology of *Drosophila* was initiated by Bill Pack and Martin Heisenberg of Tübingen.

In 1977, the Banbury Conference Center was established, using 45 acres of land, banner house, and estate building donated to the Lab by Charles Robertson. Conversion of the former seven-car garage into a meeting center gave CSH Laboratory an ideal site for holding lecture courses in neurobiology. In 1979, I organized the first lecture course at Banbury called "The Neurobiology of Behavior," which I gave with the help of Keir Pearson, Larry Squire, Fernando Nottebaum, and John Koester. Over the years, this course metamorphosed into a course on the Molecular Biology of Learning and Memory which continues to this day, now led by Kelsey Martin, an Associate Professor at the University of California, Los Angeles, whom I had the good fortune to have as a postdoctoral student in my laboratory a decade ago.

Soon there were courses on the synapse, on neural development, on new neuroanatomical methods, on pain, on the organization of the CNS, and on computational neural science, several areas that contributed to the emergence of a new science of mind.

From the outset, these courses had a worldwide influence with 15–20% of the students coming from Europe. The initial effort culminated with a major Cold Spring Harbor Symposium on The Synapse in 1975 that attracted 240 participants who listened to 60 lectures.

With both lab and lecture courses going well and covering progressively more areas of neurobiology, Jim thought that the time had come for the Lab to have its own in-house program in neural science. So in November 1978, Jim invited Birgit Zipser, who had been a student in the Leech course in 1975, to initiate the first year-round program in neurobiology at CSH Laboratory. Zipser soon initiated a classic collaboration with Ron McKay, a molecular biologist, who had been working at the Lab making monoclonal antibodies against oncogenic proteins encoded by tumor viruses. Working together, Birgit and Ron injected nerve cords from the leech into mice and discovered 40 different monoclonal antibodies, each recognizing a different set of leech neurons! A new molecular era in developmental neurobiology was emerging.

The long-term goal of neurobiology is to understand behavior, the mechanisms whereby we perceive, move, think, and remember. Until 1980, these problems were primarily being addressed at the level of individual nerve cells by examining three key questions: How does the nervous system develop? Once developed, how do nerve cells in the brain communicate with one another? How do the interactions in this complex system lead to memory and learning, and, most mysteriously, to our awareness of self?

With the results of Zipser and McKay at hand and the ability to clone genes becoming a routine laboratory exercise, Jim began to consider the possibility that all of these questions might soon be addressed on the molecular level. He therefore organized the 1983 Cold Spring Harbor Symposium on the topic Molecular Neurobiology and asked me to summarize the Symposium. The resulting meeting proved to be important historically for two reasons. The talks presented at that meeting highlighted by Shigetata Numa's delineation of the complete sequence of the four subunits of the nicotinic acetylcholine receptor illustrated that molecular biology has allowed neurobiologists to address a family of questions central to understanding how neurons signal. In addition, the talks revealed new and unexpected relationships between neurobiology and the rest of biology.

Stimulated importantly by that meeting, cellular neurobiology turned increasingly toward molecular biology and, in the subsequent years, neurobiology helped unexpectedly to uncover a deep and intellectually satisfying unity within all of biology. In this way, neurobiology contributed to the delineation of a general plan for cell function by illustrating how the cells of the nervous system are governed by variations on universal themes.

Molecular biology also opened the possibility for understanding the genetic diseases that devastate the brain, spinal cord, and muscle: muscular dystrophy, Huntington's disease, and Alzheimer's disease. As a result, a course on Neurobiology of Disease was introduced in 1984.

One of the consequences of this course is that it helped bring psychiatry, once as isolated from medicine as neurobiology was from biology, within the framework of modern cell and molecular biology. The techniques used to find genes involved in disorders such as Duchenne muscular dystrophy and Huntington's disease were being used for psychiatric disorders. However, psychiatric disorders such as schizophrenia and depression proved complex, involving several genes interacting with environmental factors. It was not at all easy to find them. Only recently have a few genes been isolated, and it now has been possible to use the transgenic methodologies pioneered in part at the Lab to study these genes in worms, flies, and mice so as to develop animal models of mental disorders. Once we begin to understand the functioning of proteins involved in psychiatric disorders, it may be possible to develop better pharmacological treatments which are much needed for mental illness.

In 1985, Jim realized it was time to develop, expand, and form a major in-house program in neuroscience on campus. He formed a committee consisting of Charles Stevens, John Klingenstein, William Robertson, and me. With the help of the Klingenstein Foundation and the Howard Hughes Medical Institute, Jim completed a new building in 1990 timed to coincide with the 100th anniversary of Cold Spring Harbor Laboratory. This was commemorated with an annual Symposium on The Brain with 285 molecular biologists, developmental biologists, neurophysiologists, cognitive neuroscientists, behavioral biologists, and model builders coming together to listen to 105 speakers. Francis Crick spoke of his interest in attention and consciousness, which revealed how far we have come from the structure of DNA! Since 1990, the in-house program has become a major force in the world of neural science. Initially, it focused primarily on the molecular biology of learning and memory through its important genetic studies of flies and mice. More recently, the Lab has combined molecular biological systems and behavioral and computational approaches to analyze complex mental functions in rodents.

In retrospect, what have the courses, meetings, symposia, and books at CSH Laboratory accomplished? They have educated the scientific community about the great emerging themes of biology, including neurobiology, and therefore created a new type of biology. In the largest sense, the courses, conferences, and symposia at the Lab provided intellectual coherence to all of biology. Until the 1970s, physicists and chemists often distinguished their fields from biology, emphasizing that biology lacked the coherence of the physical sciences. The discovery of the structure of DNA and the biology it opened up for us proved that this was no longer true. Soon, biology built on the Watson-Crick foundation helped elucidate the outline of a general plan for cell function by exploring the cell in various contexts: control of cell cycle, cell lineage, growth proliferation, transformation, oncogenesis, cell-cell interactions. In so doing, biology moved naturally to nerve cells and explored at the molecular level neurite outgrowth, synapse formation, and synapse modification by experience. As a result, we learned that regulatory genes, effector genes, and the second-messenger molecules that control functioning of target proteins arise from gene families that have characteristics in common, characteristics shared by all cells including those of the nervous system.

For the last 30 odd years, these courses, meetings, and symposia have thus delineated a unity that encompasses all of biology. The student of the brain can now ask: What underlies this unity? The unifying theme for biology is evolution, and, not surprisingly, evolutionary considerations are providing a unifying view of the wealth of information coming from the molecular analysis of living organisms. Moreover, this evolutionary perspective is proving to be an extraordinarily powerful tool in understanding protein function in the brain.

As we look toward the future, we now need to ask: How far will this unity extend? Will these principles, so helpful in understanding cell signaling and development, also lead us to a biological understanding of higher mental functions, of perception, movement, thought, and consciousness? Despite important molecular themes shared with other cells, however, the very structure and development of the neuron make it clear that the neuron is quite distinct. This problem—the question of cellular differentiation—is shared of course with other cell types. But we in neurobiology face the problem of understanding the brain's unique computational power, a power that resides in the way large numbers of cells are interconnected. Thus, if we are to come to grips with the biological underpinnings of mental processes, we shall have to go beyond the individual neuron and its immediate interactions and analyze as well the logic of information processing in large interacting systems of nerve cells that control behavior. Fortunately, important inroads even to the logic of behavior will come from molecular biology. Behavior is determined by the precise pattern of interconnections between cells. Learning, the modification of behavior by experience, is thought to result from alterations in the strength of these connections. The first inroad, the study of how connections form during development, is now well under way. This study will tell us a great deal about the underlying rules that relate different patterns of interconnections to different aspects of behavior.

To harness molecular biology and direct it to the most challenging problems confronting the behavioral and biological sciences, Jim has continued to organize at Cold Spring Harbor Laboratory a combination of molecular and behavioral approaches to study first the neural circuits in learning and memory in the fly and in the mouse and later even more complex mental problems—perception, thought, consciousness, and their disorders, autism and schizophrenia. A new terrain awaits exploration on the other side of the mountain, and with the help of Michael Wigler's pioneering work on spontaneous mutation through copy-number variations, CSH Laboratory has started and led the ascent.

Looked at in the perspective of these 30 years, we can see that biology has moved brilliantly, and this movement has been inspired by two sources. From the discovery of the structure of DNA by Watson and Crick to the delineation of the Central Dogma of the genetic code, and the early move into development, Francis Crick provided a major source of leadership. During this period, Jim Watson's influence was felt at Harvard where he directed a great lab that included Walter Gilbert and Mark Ptashne and where he wrote the textbook *Molecular Biology of the Gene* that explained the new biology to eager undergraduates, and *The Double Helix*, his autobiographical essay that explained molecular biology of the gene to the general reader. With the move to the CSH Laboratory, Jim as director of the Lab picked up where Crick left off and applied the forward-looking pedagogical style characteristic of Jim to other fields, including development, oncogenesis, immunology of the biology of the brain, the genomes of experimental animals and people, and the molecular biology of disease. Thus, as biology matured in the last third of the 20th century, one of its most significant sets of directional signals came from the Cold Spring Harbor Laboratory and its director.

NEUROSCIENCE

Getting the Point at Cold Spring Harbor

RON MCKAY

***

M Y INTEREST IN NEUROSCIENCE HAD BEEN STIMULATED while still in Edinburgh as a graduate student with Ed Southern, where nucleic acid hybridization was at the center of our world. It was there that I also first learned of the very different world being defined by Steve Kuffler, David Hubel, and Torsten Wiesel. They showed that a precise map links the axons projecting from the retina to the thalamus and cortex. There were two possible explanations for this map: It either was created by visual activity or already existed when visual activity first occurred. These two alternatives capture the central opposition in neuroscience. The Harvard professor E.O. Wilson, famous for his contributions to biology and his cognitive dissonance with Jim Watson, once remarked "philosophy consists of failed models of the brain" (see http://seedmagazine.com/news/2006/10/eo_wilson_daniel_dennett.php, search "failed models"). This quip captures the importance of the antagonism between nature and nurture in human thought. Hubel and Wiesel are recognized for their analysis of how visual activity molds the organization of the visual system, and, along with CalTech professor Roger Sperry, they were awarded the Nobel Prize

***

RON MCKAY received a B.Sc. in 1971 and a Ph.D. in 1974 from the University of Edinburgh. He received postdoctoral training at the University of Oxford working with Walter Bodmer and, in 1978, he became a Senior Staff Investigator at Cold Spring Harbor Laboratory where he worked on both SV40 T antigen and the molecular organization of the nervous system. He joined the MIT faculty in 1984, and in 1993, he moved to NINDS at the NIH as chief of the Laboratory of Molecular Biology.

in 1981. Sperry, working on the visual projection in goldfish, realized that the retinotectal map was present as soon as the fish showed visually guided behaviors. Sperry interpreted this as evidence for a sophisticated biological substrate that optimized the first sensory experience (Sperry 1963).[1] Nature creates a precise map for nurture to act on.

I moved from Edinburgh to Oxford where there was a strong tradition in structural biology and Rodney Porter was revered for his studies on immunoglobulin structure. At this time, Cesar Milstein and Georges Köhler in Cambridge showed that monoclonal antibodies could be obtained from hybrid cells generated by fusing B cells with a plasmacytoma cell line (Köhler and Milstein 1975).[2] As a postdoc with Walter Bodmer in Oxford, I realized that Sperry's speculation could be tested by making hybridomas. The power of the hybridoma approach is derived from a central feature of the immune response: One B cell produces only one immunoglobulin. This overcomes the major technical problem inherent in Sperry's proposal: the mysterious nature of the chemicals responsible for building the brain as the entire nervous system could now be used to stimulate an immune response and monoclonal antibodies screened one by one. If Sperry was right, the different antibodies would have very different anatomical distributions on different neurons.

Within a few days of arriving at Cold Spring Harbor in the fall of 1978, I fell into conversation with David Zipser walking on Bungtown Road, which runs from Route 25A through "the Lab" to the sand spit that stretches into the harbor and gives a view of Connecticut across Long Island Sound. David was a Harvard bacterial geneticist who later turned to theoretical neurobiology. Sitting in Blackford Hall, in a room opposite the cafeteria that is now decorated with images of scientific luminaries, I explained to David my attempts to test Sperry's idea by making antibodies against the rat nervous system. David's wife Birgit worked on freshwater leeches and he suggested this might be a better system to test out this slightly wacky idea. The simple repetitive structure of the leech nerve cord proved an ideal test of the general truth of Sperry's suggestion.

Birgit Zipser worked in the splendor of Jones Lab at Cold Spring Harbor. This is a 19th century purpose-built lab, impractical for molecular biology, but beautifully set on the edge of the Harbor and perfect for leech neuroanatomy. It was here that Birgit and I tested many antibodies against leech neurons. The leech nervous system is a string of beads. Each bead contains 200 neurons and is connected to the next by their axons. So the question was do the antibodies recognize highly specific subsets of cells in each bead? On one occasion, Jim Watson joined a group of us young scientists at lunch in Blackford. He asked what was the point of these experiments with leeches. I responded "to find out if the nervous system is made of shades of gray or millions of different colors." Jim was well aware of the power of antibodies because Klaus Weber, Bob Goldman, Keith Burridge and others at

Cold Spring Harbor had used antibodies against cytoskeletal proteins to transform cell biology into a molecular science. After the lunch conversation at Blackford, I was anxious that I had said something ridiculous. I need not have worried.

Jim often turns up with an open-ended or pointed question. One evening, when I was working late building a model of DNA, I realized someone was watching. Jim was leaning against the doorpost. "That's been done, Ron" was his simple statement. A few days after the Blackford exchange, Jim turned up in Jones Lab and asked to see the leech data. He sat at the microscope and looked at the antibodies illuminating new patterns of neurons, one color after the other. Jim can be economical with words. On this occasion, he looked up from the microscope with a blank stare. "Gee," he said as he left. But Jim had got the point and rapidly sponsored this project. The antibody experiment suggested that there were thousands of different chemical patterns, every neuron in the leech was different. The initial report is presented here (Zipser and McKay 1981; see also Zipser, p. 151).[3] Georges Köhler cited our work in his presentation on receiving the Nobel Prize (Köhler 1984).[4] With Susan Hockfield, the work on the leech was extended and we were now also in a position to study the interface between nature and nurture in the developing vertebrate visual system (Hockfield et al. 1983).[5]

In 1983, the Cold Spring Harbor Symposium was devoted to the new field of Molecular Neurobiology. The meeting included papers from different perspectives and it was the first time that many of the participants shared the same stage. Bernard Katz and his colleagues pioneered our understanding of neurotransmission by showing that the quantal effects of neurotransmitters were short-lived compared to the endplate potential that causes muscle depolarization. In 1982 and 1983, a series of papers documented the first clones of the nicotinic acetyl choline receptor isolated from the neuromuscular junction (NMJ) of electric ray, *Torpedo californica*. Fifteen of the papers at the symposium were focused on the new molecular biology of the neuromuscular junction, presaging two decades of molecular analysis of synaptic structure and function. Other papers reported molecular tags for voltage-sensitive sodium channels and potassium channels. The authors included the same Norman Davidson who had been among the first to understand the kinetics of nucleic acid hybridization and was a distant hero from graduate work in Edinburgh.

The intellectual atmosphere was classic Cold Spring Harbor. Outside Blackford, Jim told David Potter, then Chairman of Harvard Neurobiology, that there was no longer any need for his Department. It was politically inappropriate and perhaps not literally true, but it was obvious that understanding the nervous system had become a problem in molecular biology. Now we are used to mice lacking genes coding for neural proteins, but the first mice carrying targeted gene alterations were not reported till 1986. Despite the huge interest in analyzing specific genes in well-established neuroscience models, the antigens we had defined in the nervous system pointed to another path. After the Symposium, Jim walked into the lab and asked a classic Watson question: "What would it take to compete with Corey Goodman?" I talked with Corey for the first time the evening before we presented at the Symposium. Corey, Nick Spitzer, and Mike Bate were studying how axons reached their targets in the grasshopper (Bate et al. 1981).[6] At the Symposium, Corey suggested that a combination of molecular biology and genetics in *Drosophila* was the way to move forward with the "labeled pathways" hypothesis, their updated version of Sperry's world. In 1983, we specifically reported antigens that identify growth cones on subsets of growing axons (McKay et al. 1983).[7] Sperry was right but our data also showed "that neurons do not differ in a single molecular characteristic, such as neurotransmitter, but in many different molecular features" (p. 608).[7] The path ahead was no longer focused solely on axon guidance. No doubt I maintained an outer calm in responding to Jim's question, but our data on the visual system had me hooked on the mammalian brain, and there was no way I was going to work on flies. I can recall wondering if I was making a convincing explanation of my new interest in the developmental origins of neuronal diversity in mammals.

Many of Jim's peers from the origins of Molecular Biology were involved in ambitious new projects that left bacteria far behind. Seymour Benzer was working on flies. Max Delbrück was working on a light-sensitive plant. Sydney Brenner was mapping the wiring diagram of the nervous system in nematodes (White et al. 1983).[8] The mantra of the time was simple genetic models. I was planning a different path, following Barbara McClintock's adumbration that "One must have the time to look, the patience to hear what the material has to say to you. Above all, one must have a feeling for the organism."

In starting this new quest, we were already in possession of a key fact. Seymour and his colleague Don Ready had published an important paper showing that the different cells in the fly's eye were all derived from a common precursor (Ready et al. 1975).[9] This precursor cell became a major focus of our interest as I describe in the essay accompanying the Hockfield and McKay (1985) paper (see also McKay, p. 157).[10]

I left Cold Spring Harbor in 1984 and with Kristen Frederiksen moved to a new home in Bethesda and a position at the National Institutes of Health. There we continued what many thought was the mad pursuit of mammalian neural stem cells. Using the Rat401 monoclonal antibody, generated at Cold Spring Harbor, we isolated the nestin gene, a new intermediate filament protein transiently expressed in central nervous system precursors and in adult stem cells. Soon we were reporting efficient strategies for generating functional neurons and glia from fetal and embryonic stem cells.

Our understanding of life is governed by the courage to promote simple insights. When Seymour moved to the California Institute of Technology, he worked with Sperry as he refocused on the behavior of flies. In the opening of his first paper on this subject, Seymour writes, "Complex as it is, much of the vast network of cellular functions has been successfully dissected, on a microscopic scale, by the use of mutants in which one element is altered at a time" (Benzer 1967).[11] A few years later with his colleague Ron Konopka, they state "the fact that a period of the rhythm under constant conditions is altered implies that the mutations are affecting a basic oscillator" (Konopka and Benzer 1971).[12] By 1994 at Cold Spring Harbor, the influence of Seymour Benzer was still felt in the work of Jerry Yin and Tim Tully exploring the conserved pathways that control memory. In this paper, mutations in the cAMP-dependent transcription factor CREB were shown to block one of two simple learning paradigms in *Drosophila* (Yin et al. 1994; see also Tully, p. 163).[13] Here is the same logic so powerfully used by the phage and splicing groups, precise forward genetics. In this case, unraveling a basic aspect of the biology of synapses.

All over the world, there are now rooms full of carefully recorded fruit flies, fish, and mice that are studied to understand the molecular biology of the nervous system. But many of the most interesting features of neuroscience are found in animals that cannot be bred in the large numbers needed for classical genetics. Of course there are many ways to understand the relationship between genes and the environment. E.O. Wilson has proposed what amounts to a biosphere genome project to save bio-diversity on the planet (see http://www.ted.com/index.php/talks/view/id/83). There are increasing numbers of laboratories using human stem cells to do forward genetics in man, to link genotype and phenotype. The molecular and cellular biology of the nervous system is now a global enterprise.

When I was at the Massachusetts Institute of Technology, Jim had asked me a typical Watsonian question: "Who do you talk to?" I was too naïve to know the right answer. But for many of us, in the bright sun of Cold Spring Harbor summers, we learned more than scientific facts. After moving to NIH, I eventually realized what Jim was saying. The interface between nature and nurture illuminated by new discoveries on genes or stem cells must be constantly "discussed" with Professors, Politicians, and Pundits. A hazardous enterprise, but an essential intellectual contribution to culture and an enduring characteristic of the dialogue at Cold Spring Harbor.

Notes and References

1. Sperry R.W. 1963. Chemoaffinity in the orderly growth of nerve fiber patterns and connections. *Proc. Natl. Acad. Sci.* **50:** 703—710.

2. Köhler G. and Milstein C. 1975. Continuous cultures of fused cells secreting antibody of predefined specificity. *Nature* **256:** 495–497.

3. Zipser B. and McKay R. 1981. Monoclonal antibodies distinguish identifiable sets of neurons in the leech. *Nature* **289:** 549–554.

4. Köhler G.J.F. Derivation and diversification of monoclonal antibodies. Nobel Lecture, 8 December 1984.

5. Hockfield S., McKay R.D., Hendry S.H., and Jones EG. 1983. A surface antigen that identifies ocular dominance columns in the visual cortex and laminar features of the lateral geniculate nucleus. *Cold Spring Harbor Symp. Quant. Biol.* **48(Pt2):** 877–889.

6. Bate M., Goodman C.S., and Spitzer N.C. 1981. Embryonic development of identified neurons: Segment-specific differences in the H cell homologues. *J. Neurosci.* **1:** 103–106.

7. McKay R.D., Hockfield S., Johansen J., and Frederiksen K. 1983. The molecular organization of the leech nervous system. *Cold Spring Harb. Symp. Quant. Biol.* **48 (Pt2):** 599–610.

8. White J.G., Southgate E., Thomson J.N., and Brenner S. 1983. Factors that determine connectivity in the nervous system of *Caenorhabditis elegans*. *Cold Spring Harbor Symp. Quant. Biol.* **48(Pt2):** 633–640.

9. Ready D.F., Hanson T.E., and Benzer S. 1976. Development of the *Drosophila* retina, a neurocrystalline lattice. *Dev. Biol.* **53:** 217–240.

10. Hockfield S. and McKay R.D.G. 1985. Identification of major cell classes in the developing mammalian nervous system. *J. Neurosci.* **5:** 3310–3328.

11. Benzer S. 1967. Behavioral mutants of *Drosophila* isolated by countercurrent distribution. *Proc. Natl. Acad. Sci.* **58:** 1112–1119.

12. Konopka R.J. and Benzer S. 1971. Clock mutants of *Drosophila melanogaster*. *Proc. Natl. Acad. Sci.* **68:** 2112–2116.

13. Yin J.C.P., Wallach J.S., Delvecchio M., Wilder E.L., Zhou H., Quinn W.G., and Tully T. 1994. Induction of a dominant-negative Creb transgene specifically blocks long-term-memory in *Drosophila*. *Cell* **79:** 49–58.

Zipser B. and McKay R. 1981. **Monoclonal antibodies distinguish identifiable neurones in the leech.** *Nature* **289:** 549–554. (Reprinted, with permission, from Macmillan Publishers Ltd ©1981.)

THE SPECIFICITY OF NEURONAL CONNECTIONS WAS POSTULATED by Roger Sperry[1] "to be the result of neurons connecting to one another via complementary molecules." The idea that chemoaffinity could explain neuronal interactions was both attractive and influential, but there was no simple assay for the neuron-specific, low-abundance molecules postulated to serve in neuron-neuron recognition. However, the development of hybridoma technology by Kohler and Milstein,[2] brought from England to Cold Spring Harbor by Ron McKay,[3] offered a method of detecting such molecules without the need for prior antigen purification. David Zipser[4] and I realized that because hybridomas are immortalized cells, with each clone secreting just one antibody, highly specific immunological probes could now be generated against rare neural epitopes by immunizing mice with a crude mixture of brain antigens. The leech central nervous system with its large identifiable neurons seemed to be an ideal system to screen for epitopes expressed by single neurons or sets of neurons in the search for neural recognition molecules. In addition, leech antigens, particularly the glycans, were highly immunogenic, in line with the hypothesis that crucial biological roles of glycans are often mediated by unusual sugar sequences that may be species-specific.[5]

The fertile reward of immunizing mice with a crude mixture of leech brain antigens was a wealth of antibodies recognizing different types of neurons and glial cells during their various developmental stages. These antibodies have been used for staining, experimental manipulations, and antigen purification, followed by gene cloning and mutations. Our pioneering work has led to an explosion of information

BIRGIT ZIPSER, Cold Spring Harbor's first appointment in Neurobiology, joined the Laboratory in 1978 from Downstate Medical College Brooklyn, where she was a Research Assistant Professor. She worked in the renovated Jones Laboratory, which for years previously had been used for summer courses in ecology and biology taught to local schoolchildren. Birgit left the laboratory to take a position at the National Institutes of Health in Bethesda and is now a Professor in the College of Natural Science, Department of Physiology, at Michigan State University. zipserb@msu.edu

on the development of different neuronal systems across many animal species.

It quickly became clear, as in much of molecular biology, that the molecules were not specific in a simple way; there was no "one molecule–one connection" relationship. Consequently, to explain our subsequent data, we abandoned Sperry's chemoaffinity hypothesis for the differential adhesion hypothesis put forward by Steinberg[6] and Steinberg and Takeichi.[7] Focusing in our subsequent studies on two types of glycoepitopes, we promoted the leech to a glycobiological model system: a constitutive mannosidic epitope, shared by all sensory afferent neurons, and late-appearing, galactosidic epitopes. These developmentally regulated, mature galactose markers are expressed by subsets of these same afferents and, most importantly, correlate with different sensory modalities.[8] All of these glycoepitopes are modifications of leech CAM and Tractin, members of the neural cell adhesion molecule (NCAM) and L1 families of proteins.[9] The constitutive polymannopyranose[10] mediates axonal sprouting of afferents,[11] and subsequently, the appearance of the developmentally regulated mature galactosidic epitopes opposes axonal sprouting and mediates the patterning of their respective sensory subsets into laminae.[12] The differential adhesion hypothesis of Steinberg and Takeichi explains how self-similar afferents, with the same galactose marker, drive segregation into different laminae.[13] Formation of laminae is a common principle of target region organization.

In the vertebrate brain, sensory laminae also are delineated by modality-specific neutral glycans.[14-16] However, injury incident to culturing dissociated sensory neu-

Birgit Zipser, circa 1982. *(Photograph by Joan James, courtesy of Cold Spring Harbor Laboratory Archives.)*

Ron McKay, 1996. *(Photograph by Marléna Emmons, courtesy of Cold Spring Harbor Laboratory Archives.)*

rons led to the replacement of their mature modality-specific epitopes by Lewis x, a glycoepitope common to fetal neurons.[14] Likewise, mutational studies of glycosyltransferases done in vitro[17] elucidate glycosylation pathways but not glycan function in the animal. Gene-targeting studies in vivo have been foiled by lethality[18] or the apparent lack of a phenotype[19] that was not further examined for subtle neurological deficits which can arise in mice growing into adulthood without obvious abnormalities.[20] In the developing vertebrate nervous system, most of the functional information so far is only available for negatively charged glycans[21]—polysialic acid, the Lewis x and HNK-1 epitopes, and GAG chains—because they are more immunogenic, easier to isolate, and less dependent on the physiological state of the cell.

In retrospect, the embryonic leech was a good place to start analyzing the neurobiological function of neutral glycans because its intact central nervous system is readily amenable to direct experimental manipulations and live imaging.[22] Illuminating the neural function of these glycans in vertebrates requires a polyvalent approach, including, for example, the topographical surveying of the outer surfaces of cells and mapping of signaling pathways between molecules on the outer cell surfaces and nucleus. A critical first step in identifying candidate glycans functioning in neural connectivity will be to target glycosyltransferases expressed in subregions or specific cell types of the brain and analyze the mutants for neurological deficits using biophysical methods and sophisticated behavioral paradigms.

Notes and References

1. Sperry R.W. 1963. Chemoaffinity in the orderly growth of nerve fibre patterns and connections. *Proc. Natl. Acad. Sci.* **50:** 703–710.

2. Kohler G. and Milstein C. 1975. Continuous cultures of fused cells secreting antibodies of predicted specificity. *Nature* **256:** 495–497.

3. Ron McKay came to Cold Spring Harbor as a postdoctoral fellow from the Oxford laboratory of Walter Bodmer. His major project—the biochemical functions of

SV40-encoded T antigen—was carried out in the James Laboratory. But the collaboration that was to permanently change his scientific life was with Birgit Zipser, who worked down the hill in the Jones Laboratory. Ron left Cold Spring Harbor in 1980 for a position at the National Institutes of Health.

4. David Zipser, a prokaryotic geneticist, joined Cold Spring Harbor from Columbia University to help the Laboratory reestablish genetics as a major focus. During his 12 years at the Laboratory, David's interest shifted to computational models of brain function, a topic that he continued after moving in 1982 to the University of California, La Jolla.

5. Varki A. 1993. Biological roles of oligosaccharides: All the theories are correct. *Glycobiology* **3:** 97–130.

6. Steinberg M.S. 1970. Does differential adhesion govern self-assembly processes in histogenesis? Equilibrium configurations and the emergence of a hierarchy among populations of embryonal cells. *J. Exp. Zool.* **173:** 395–434.

7. Steinberg M.S. and Takeichi M. 1994. Experimental specification of cell sorting, tissue spreading and specific spatial patterning by quantitative differences in cadherin expression. *Proc. Natl. Acad. Sci.* **91:** 206–209.

8. Zipser K., Erhardt M., Song J., Cole R.N., and Zipser B. 1994. Distribution of carbohydrate epitopes among disjoint subsets of leech sensory afferent neurons. *J. Neurosci.* **14:** 4481–4493.

9. Jie C.Y., Xu D., Wang B., Zipser J., Jellies K., Johansen M., and Johansen J. 2000. Posttranslational processing and differential glycosylation of Tractin, an Ig-superfamily member involved in regulation of neuronal outgrowth. *Biochim. Biophys. Acta* **1479:** 1–14.

10. Huang L. Hollingsworth R., Husain R., and Zipser B. 2002. Separation and characterization of cell type-specific poly-b (1,4) linked mannopyranose from the leech species *Hirudo medicinalis. Glycobiology* **12:** 679.

11. Zipser B., Morell R., and Bajt M.L. 1989. Defasciculation as a neuron pathfinding strategy: Involvement of a specific glycoprotein. *Neuron* **3:** 621–630.

12. Song J. and Zipser B. 1995. Targeting of neural subsets mediated by their sequentially-expressed carbonydrate markers. *Neuron* **14:** 537–547.

13. Tai M.-H. and Zipser B. 2002. Sequential steps of carbohydrate signaling mediate sensory afferent differentiation. *J. Neurocytol.* **31:** 743–754.

14. Dodd J. and Jessell T.M. 1985. Lactoseries carbohydrates specify subsets of dorsal root ganglions projecting to superficial dorsal horn of rat spinal ganglia. *J. Neurosci.* **5:** 3278–3294.

15. Key B. and Akerson R.H. 1991. Delineation of olfactory pathways in frog nervous system by unique glycoconjugates or N-CAM isoforms. *Neuron* **6:** 381–396.

16. Pays L. and Schwarting G. 2000. Gal-NNCAM is a differentially-expressed marker for mature sensory neurons in the rat olfactory system. *J. Neurobiol.* **43:** 173–185.

17. Patnaik S.K. and Stanley P. 2006. Lectin-resistant CHO glycosylation mutants. *Methods Enzymol.* **416:** 159–182.

18. Ioffe E. and Stanley P. 1994. Mice lacking *N*-acetylglycosaminyltransferase I activity die at mid-gestation, revealing an essential role for complex or hybrid *N*-linked carbohydrates. *Proc. Natl. Acad. Sci.* **91:** 272–232.

19. Bhattacharyya R., Bhaumik M., Raju S., and Stanley P. 2002. Truncated, inactive N-acetylglucosaminyltransferase III (GlcNAc-TIII) induces neurological and other traits absent in mice that lack GlcNAc-TIII. *J. Biol. Chem.* **277:** 26300–26309.

20. Tsien J.Z., Huerta P.T., and Tonegawa S. 1996. The essential role of hippocampal CA1 NMDA receptor-dependent synaptic plasticity in spatial memory. *Cell* **87:** 1147–1148.

21. Matani P.M., Sharrow M., and Tiemeyer M. 2007. Ligand, modulatory, and co-receptor functions of neural glycans. *Front. Biosci.* **12:** 3852–3879.

22. Baker M.W., Kauffman B., Macagno E.R., and Zipser B. 2003. In vivo dynamics of the CNS sensory arbor formation: A time-lapse study in the embryonic leech. *J. Neurobiol.* **56:** 41–53.

Monoclonal antibodies distinguish identifiable neurones in the leech

Birgit Zipser & Ronald McKay

Cold Spring Harbor Laboratory, PO Box 100, Cold Spring Harbor, New York 11724

Monoclonal antibodies were isolated by screening 475 hybridomas obtained from mice immunized with whole leech nerve cords. The majority (about 300) reacted with leech nervous tissue, but only about 40 made antibodies that identified single kinds or small sets of cells. Twenty of the antibodies which react with specific neurones were studied in greater detail and are described here. They include antibodies against identified sensory neurones and motor neurones as well as against numerous unidentified cells.

THE ability of a nervous system to generate coherent behaviour is dependent on a vast number of precise connections between neurones. Virtually nothing is known about the molecular mechanisms which generate these connections, but all tenable hypotheses ultimately postulate the existence of molecules, differing in kind or quantity from cell to cell, which mediate the necessary recognition. The problem of neural specificity becomes apparent when one considers that neurones typically receive inputs from and send outputs to whole sets of other cells each of which in turn has its own extremely complex characteristic connection pattern. Connections are established during development but the degree to which marker molecules may be present in adults is unknown, although regeneration studies in adults[1] imply long-term persistence of at least some molecular specificity. It has long been known that there are many chemical differences between neurones, involving such functions as transmitter synthesis. Recent studies on the identification and localization of neuronal peptides[2] have greatly widened the scope of detectable molecular variation among nerve cells. The relationship of this chemical variation to the mechanism of connection is still unknown. Progress in this area will require the development of ways of relating chemical specificity to unique connectivity. Because, by hypothesis, the markers of interest will differ in each neurone with different connections, these molecules must somehow be identified in structured material where the same neurone can be easily identified in repeated experiments. Such a simple system is required not only for the identification of specific neuronal markers but also for analysing the rules which establish neuronal connections.

As a step towards this goal, we have investigated whether individual cells and sub-networks of the relatively simple leech nervous system could be distinguished by specific antibodies. As the putative antigens could not, in principle, be purified *a priori*, we used the technique of hybridoma formation[3], which allows the isolation of lymphocyte clones secreting antibodies specific for individual molecules even though a chemical mixture of antigens was used for immunization. Monoclonal antibodies were obtained from lymphocytes of mice immunized with the entire isolated nervous system of the leech and screened on intact ganglia. The results of our study, reported here, give a

clear answer to the question of the existence of chemical markers in individual cells and sub-networks. These specific markers are surprisingly abundant, and extrapolation from our initial sample makes it feasible that every cell has one or more chemical markers shared by only small subsets of neurones. Although we know little about the exact cellular locations of these markers, many are present in all parts of the cell, including long axonal projections and neural terminals. Indeed, the situation seems quite analogous to a colour-coded electric cable

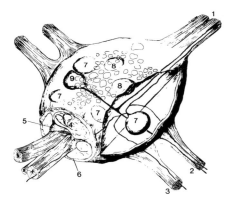

Fig. 1 Diagram of a midbody ganglion. (1) Connective, (2) anterior root, (3) posterior root, (4) neuronal cell bodies in glial packages, (5) the beginning of the neuropile where synapses occur, (6) capsule, (7) two pairs of bilaterally symmetrical pressure cells, (8) the pair of large Retzius cells are shown in the background of several hundred neuronal cell bodies, (9) one of the two lateral penile evertor cells (the other has been dissected away). Each cell body has a characteristic location and number of axons. Note that the penile evertor motor neurone projects into contralateral roots and the sensory pressure cell has ipsilateral projections.

0028-0836/81/070549-06$01.00

Hockfield S. and McKay R.D.G. 1985. **Identification of major cell classes in the developing mammalian nervous system.** *J. Neurosci.* **5:** 3310–3328. (Reprinted, with permission, from Society for Neuroscience © 1985.)

TECHNIQUES BASED ON THE CHEMISTRY of nucleic acids and the manipulation of viruses were not the only tools used to good effect at Cold Spring Harbor. Hybridoma technology, founded on the landmark research of George Köhler and Cesar Milstein at the MRC Laboratory for Molecular Biology at Cambridge, was rapidly adopted at CSH Laboratory. Köhler and Milstein (1975)[1] had shown that antibodies with their exquisite specificity could be derived from clonal cell lines, and these monoclonal antibodies opened up a new level of molecular analysis. In 1981, a small meeting on the use of hybridoma technology in neuroscience was held at the Banbury Conference Center. This meeting had an atmosphere of excitement, typical when a breaking technology is discussed at Banbury.

By 1983, the power of monoclonal antibodies was evident in the Cold Spring Harbor Symposium on Molecular Neurobiology. It was clear from the Symposium that the new field of molecular neuroscience would need to combine multiple traditional and new disciplines, and in the next decade, model systems with powerful genetics and recombinant DNA methods were developed. These studies focused mainly on the molecules thought to generate the unique qualities of the adult brain, and of these, ligand-gated receptors and ion channels had a special iconic quality. Hybridomas were employed by many of the speakers at the 1983 Symposium, including Josh Sanes, Martin Raff, Melitta Schachner, Art Lander, Tom Jessell, Lou Reichardt, Corey Goodman, and the Cold Spring Harbor neuroscientists—Birgit Zipser, Sue Hockfield, and myself.

The 1985 paper reprinted here marked the next phase, where the focus was not on molecules but cells. We knew that a precisely timed developmental program generated different neurons in the mammalian brain, as the cells left the proliferating

---

RON McKAY (*See biographical footnote on p. 145.*)

Sue Hockfield, circa 1982. *(Courtesy of Cold Spring Harbor Laboratory Archives.)*

state and became postmitotic. So we set out to find the dividing cells that generate neurons, by making antibodies that would identify all of the cell types in the developing brain. As we noted in the introduction to this paper, "these reagents could then be used to study cellular mechanisms of development in the early neural tube which give rise to the differentiated adult vertebrate brain." Seymour Benzer and his colleagues at the California Institute of Technology were carrying out experiments using antibodies to study the developing retinal imaginal disc in the fly (Fujita et al. 1982; Zipursky 1984; Banerjee et al. 1987a,b).[2–5] We wanted to pose a similar general question: How does the molecular diversity of vertebrate neurons arise?

Interestingly, molecular studies of mouse development were started at this time. For example, Alex Joyner with Bob Tjian and Gail Martin as co-authors published a paper in *Nature* defining conserved homeobox genes in mouse and man (Joyner et al. 1985),[6] and Peter Gruss published his first paper on Pax genes in 1988 (Deutsch et al. 1988).[7] In contrast to this research, from the beginning, our focus was on the integrated activity of many genes and we achieved this by studying the cells that make the brain. Pat Levitt working with Pasko Rakic had shown that neuronal and glial precursors coexisted during the development of the monkey brain (Levitt et al. 1981).[8] We were interested in studying a time in development when we could obtain a complete classification of the founding cells in the central nervous system (CNS) and understanding how they interact to generate the complex adult tissue.

To identify these cells, we made hybridomas against the spinal cord at a time when neurons were being generated. The precise timing of neuron formation in the rat spinal cord had been determined using radioactive precursors for DNA and autoradiography (Nornes and Das 1974).[9] So we dissected the spinal cord from rat embryos on day 15 of development, just at the end of the period when most neurons are formed, and immunized mice with homogenized tissue. We fixed the spinal cord with 4% paraformaldehyde before immunizing the mice, to optimize the chance of getting antibodies against antigenic epitopes present in tissue that had been fixed with paraformaldehyde for immunohistochemistry. As we are reminded every time we get the flu, it takes time for our immune system to get going. We had become expert at timing the immune response in our mice, and, at a time when the spleen contained many B cells responding to the antigen, we used polyethylene glycol to make hybridoma cell lines by fusing the spleen cells with an immortal plasmacytoma cell line generated at the National Institutes of Health. We immunized several mice and carried out multiple cell fusions. Another important technical advance that we called "hitting the jackpot" unfortunately was not under tight control, but one out of every two or three mice would generate many hybridoma cell lines.

The hybridomas were screened by immunohistochemistry using sections of the brains of rat embryos that had been perfusion-fixed. Perfusing an embryonic day-12 rat embryo through the heart requires a steady hand. The benefit of this meticulous approach can still be clearly seen by the high quality of the light and electron microscopy in the paper. There are several aspects of these results worth noting. Figures 1 to 3 define the most prevalent cell types in the developing CNS at this critical time when different neurons are born. These cells included neurons, vascular cells, and an epithelial precursor cell. Most of the paper is concerned with the antibody Rat401 that recognizes a high-molecular-weight protein in a transient epithelial precursor cell. As Pat Levitt said when looking at this staining pattern, "it goes away." Indeed it does.

Figures 4 to 7 show the morphology of the cells expressing the antigenic epitope recognized by the Rat401 antibody. The antigen first appears at the outer surface of epithelial cells and then fills these cells, showing that they stretch across the neural plate in the period before neurons are born. However, once neurons are present, the Rat401 antigen defines an elongated cell with a distinct morphology, the radial glial cell. The distribution of the Rat401 antigen in the newborn neurons can be seen by comparing Figures 6A and 9A. These data showed that the Rat401 antigen was expressed in the precursor cells throughout the developing brain.

The epitope bound by the Rat401 antibody was defined as a new intermediate filament protein, nestin (Lendahl et al. 1990),[10] and additional experiments proved that nestin-positive cells from the fetal and adult brain give rise to neurons and glia in vivo and in cell culture (Frederiksen and McKay 1988; Cattaneo and McKay 1990; McKay et al. 1990; Reynolds and Weiss 1992).[11–14] This precise definition of neural stem cells also allowed the efficient derivation of neural stem cells and functional neurons from embryonic stem cells (Okabe et al. 1996).[15] Antibodies against nestin and the regulatory regions from the nestin gene are now used all over the world to identify and manipulate neural precursor cells. The idea of a neural stem cell is explicitly stated in the 1985 report which was clearly an important step in the identification of mammalian somatic stem cells.

Notes and References

1. Köhler G. and Milstein C. 1975. Continuous cultures of fused cells secreting antibody of predefined specificity. *Nature* **256:** 495–497.

2. Fujita S.C., Zipursky S.L., Benzer S., Ferrus A., and Shotwell S.L. 1982. Monoclonal antibodies against the *Drosophila* nervous system. *Proc. Natl. Acad. Sci.* **79:** 7929–7933.

3. Zipursky S.L., Venkatesh T.R., Teplow D.B., and Benzer S. 1984. Neuronal development in the *Drosophila* retina: Monoclonal antibodies as molecular probes. *Cell* **36:** 15–26.

4. Banerjee U., Renfranz P.J., Hinton D.R., Rabin B.A., and Benzer S. 1987a. The sevenless+ protein is expressed apically in cell membranes of developing *Drosophila retina*; It is not restricted to cell R7. *Cell* **51:** 151–158.

5. Banerjee U., Renfranz P.J., Pollock J.A., and Benzer S. 1987b. Molecular characterization and expression of sevenless, a gene involved in neuronal pattern formation in the *Drosophila* eye. *Cell* **49:** 281–291.

6. Joyner A.L., Lebo R.V., Kan Y.W., Tjian R., Cox D.R., and Martin G.R. 1985. Comparative chromosome mapping of a conserved homoeobox region in mouse and human. *Nature* **314:** 173–175.

7. Deutsch U., Dressler G.R., and Gruss P. 1988. Pax 1, a member of a paired box homologous murine gene family, is expressed in segmented structures during development. *Cell* **53:** 617–625.

8. Levitt P., Cooper M.L., and Rakic P. 1981. Coexistence of neuronal and glial precursor cells in the cerebral ventricular zone of the fetal monkey: An ultrastructural immunoperoxidase analysis. *J. Neurosci.* **1:** 27–39.

9. Nornes H.O. and Das G.D. 1974. Temporal pattern of neurogenesis in spinal cord of rat. I. An autoradiographic study—Time and sites of origin and migration and settling patterns of neuroblasts. *Brain Res.* **73:** 121–138.

10. Lendahl U., Zimmerman L.B., and McKay R.D. 1990. CNS stem cells express a new class of intermediate filament protein. *Cell* **60:** 585–595.

11. Frederiksen K. and McKay R.D. 1988. Proliferation and differentiation of rat neuroepithelial precursor cells in vivo. *J. Neurosci.* **8:** 1144–1151.

12. Cattaneo E. and McKay R. 1990. Proliferation and differentiation of neuronal stem cells regulated by nerve growth factor. *Nature* **347:** 762–765.

13. McKay R., Valtz N., Cunningham M., and Hayes T. 1990. Mechanisms regulating cell number and type in the mammalian central nervous system. *Cold Spring Harbor Symp. Quant. Biol.* **55:** 291–301.

14. Reynolds B.A. and Weiss S. 1992. Generation of neurons and astrocytes from isolated cells of the adult mammalian central nervous system. *Science* **255:** 1707–1710.

15. Okabe S., Forsberg-Nilsson K., Spiro A.C., Segal M., and McKay R.D. 1996. Development of neuronal precursor cells and functional postmitotic neurons from embryonic stem cells in vitro. *Mech. Dev.* **59:** 89–102.

0270-6474/85/0512-3310$02.00/0
Copyright © Society for Neuroscience
Printed in U.S.A.

The Journal of Neuroscience
Vol. 5, No. 12, pp. 3310–3328
December 1985

Identification of Major Cell Classes in the Developing Mammalian Nervous System[1]

S. HOCKFIELD[2] AND R. D. G. McKAY[3]

Cold Spring Harbor Laboratory, Cold Spring Harbor, New York 11724

Abstract

A major difficulty in studying early developmental processes and testing hypotheses of possible cellular mechanisms of development has been the inability to reproducibly identify specific cell types. We have generated monoclonal antibodies that distinguish among major cell types present during mammalian neurogenesis. These antibodies have been used to analyze the development of cellular organization in the early nervous system. Monoclonal antibody Rat-401 identifies a transient radial glial cell in the embryonic rat central nervous system (CNS) that is temporally and spatially suited to guide neuronal migration. Rat-401 also identifies a peripheral non-neuronal cell that may establish axon routes from the CNS to the periphery. Monoclonal antibody Rat-202 recognizes an antigen present in early axons, their growth cones, and filopodia, and has allowed us to follow early axons and observe the structures they contact. Two other antibodies that recognize axons demonstrate antigenically distinct phases in axon development. In addition, we report a marker for another cell class present in the developing nervous system, the endothelial cells that give rise to the CNS vasculature.

Molecular techniques have confirmed and extended earlier anatomical and physiological studies showing that the adult vertebrate central nervous system (CNS) is composed of a large number of different cell types. This large number of cell types is derived during a short period of embryonic development from a small number (approximately 10^4; unpublished observations) of morphologically homogeneous neuroepithelial cells at embryonic day 10 in the rat. In response to a mesodermal signal (Spemann, 1936) the early neuroectodermal cells proliferate and differentiate to give rise to the adult brain which contains a morphologically heterogeneous population of greater than 10^9 neurons and glial cells. Particular developmental features of the early neural tube, such as the birthdate of

neurons (Sidman, 1970) and the guiding role of radial glial cells (Rakic, 1971), are thought to have important consequences for the synaptic organization of the adult brain. Several recent studies have used cell type-specific markers to study cellular differentiation in the peripheral nervous system (PNS) (Le Douarin, 1980; Barald, 1982; Vincent and Thiery, 1984). Fewer cell type-specific markers have been described in the vertebrate CNS. Recent progress in using hybridoma technology to identify subsets of adult (Hawkes et al., 1982; McKay and Hockfield, 1982; Sternberger et al., 1982) and embryonic (Levitt, 1984) vertebrate neurons and to study glial cell origins (Raff et al., 1983) suggested to us that a similar strategy might provide reagents to identify cell types in the developing vertebrate CNS. These reagents could then be used to study cellular mechanisms of development in the early neural tube which give rise to the differentiated adult vertebrate brain.

In this paper we describe a set of reagents which identify major structures and developmental periods in the embryonic nervous system of the rat. We have used three reagents to describe the cellular organization of the embryonic nervous system and to study its development. These reagents have enabled us to identify three classes of cellular elements in the developing mammalian nervous system: axons; radial glia and early Schwann-like cells; and early endothelial cells. We have used three monoclonal antibodies to study axonal development and have shown that each antibody recognizes antigens that are expressed at distinct stages in development. One antibody labels early-growing axons and their growth cones and filopodia, enabling us to follow axons as they grow from the CNS to the periphery, whereas the two other antibodies label axons at later stages in their development. Another monoclonal antibody recognizes the early radial glial cells but does not recognize cells in the adult CNS. The transient expression of the antigen recognized by this antibody correlates with the period of neuronal proliferation and migration. This antibody also identifies a non-neuronal cell that appears to precede and predict axon pathways from the CNS into the periphery. We also describe here a marker for cells that give rise to the CNS vasculature.

Received March 6, 1985; Revised June 27, 1985;
Accepted July 2, 1985

[1] We would like to thank Elizabeth Waldvogel and Carmelita Bautista for their continued excellent technical assistance, Dr. B. Friedman for discussions of the manuscript, Mike Ockler for graphics, and Marlene Rubino for secretarial help. We thank Dr. L. Eng, Dr. B. Pruss and Dr. S. Blose for their generosity in supplying antibodies to, respectively, glial fibrillary acidic protein, intermediate filaments, and vimentin. National Institutes of Health Grants NS 18040 (S. H.) and NS 17556 (R. D. G. M.) and National Science Foundation Grant BNS 84-19240 (S. H.) supported this work.

[2] To whom correspondence should be addressed, at her present address: Section of Neuroanatomy, Yale University School of Medicine, 333 Cedar St., New Haven, CN 06510.

[3] Present address: Whitaker College and Department of Biology, E25-435, Massachusetts Institute of Technology, Cambridge, MA 02139.

Materials and Methods

In order to obtain markers for the major cell classes in the developing neural tube we generated monoclonal antibodies to fixed spinal cord from embryonic day 15 (E15) rats. Timed pregnant female rats were obtained from Taconic Animal Supply Co. and housed until the appropriate gestational age. Uteri were dissected from pregnant animals into ice-cold phosphate buffer (pH 7.4), individual embryos were removed, and the spinal cord was dissected free of other tissue (4% paraformaldehyde in 0.1 M phosphate buffer. We followed an immunization and fusion protocol that has been described previously (McKay and Hockfield, 1982). Briefly, BALB/c mice received two intraperitoneal immunizations with fixed tissue homogenized in saline and suspended in an equal volume of Freund's complete adjuvant and a final intravenous boost of unfixed tissue without adjuvant. Spleen cells from immunized mice were fused with NS-1 myeloma cells, and resulting hybrid cell lines were screened immunohistochemically

Yin J.C.P., Wallach J.S., Del Vecchio M., Wilder E.L., Zhou H., Quinn W.G., and Tully T. 1994. **Induction of a dominant negative CREB transgene specifically blocks long-term memory in Drosophila.** *Cell* **79:** 49–58. (Reprinted, with permission, from Elsevier ©1994.)

I N SEPTEMBER 1991, I ARRIVED AT MY NEW JOB in the new Beckman Building at Cold Spring Harbor Laboratory. By then, I had spent several years working on memory formation in *Drosophila*, first as a postdoctoral fellow with W.G. "Chip" Quinn at Princeton and the Massachusetts Institute of Technology and then as an assistant professor at Brandeis University. In the early days at Princeton, I had adapted the instrumental olfactory conditioning procedure of Quinn, Harris, and Benzer (Quinn et al. 1974)[1] to an experimentally more manageable classical (or Pavlovian) conditioning protocol and demonstrated that fruit flies displayed behavioral properties of Pavlovian learning and memory quite similar to those shown in other species (Tully and Quinn 1985).[2] To a geneticist, this behavioral "homology of function" suggested an evolutionarily conserved underlying molecular mechanism. My colleagues and I then turned our attention (1) to screen for single-gene mutants that disrupt learning and/or memory (Dubnau and Tully 1998; Dubnau et al. 2003)[3,4] and (2) to use the growing collection of mutants for a "genetic dissection" of learning/memory, as conceptually advanced by Seymour Benzer (1973).[5]

The new Pavlovian protocol produced robust learning, thereby allowing a more detailed study of subsequent memory formation per se. Behavioral studies of memory formation in humans and other animals have generally shown that a new memory initially decays and is easily disrupted. With time (and practice), however, early

TIM TULLY received B.S. degrees in Biology and Psychology from the University of Illinois in 1976 and a Ph.D. in Genetics in 1981. Postdoctoral training followed, first in Neurogenetics at Princeton University and then in Molecular Genetics at MIT. He joined the faculty at Cold Spring Harbor Lab in 1991 after four years on the faculty of Brandeis University. In 2007, he left CSHL to take up his current position as Chief Science Officer of Dart Neuroscience LLC. He remains an Adjunct Professor at Tsinghua University in Beijing, China, and at the National Tsing Hua University in Hsin Chu Taiwan. tully.tp@gmail.com

Tim Tully, 1994. *(Courtesy of Cold Spring Harbor Laboratory Archives.)*

decremental memory becomes "consolidated" into a stable, long-lasting memory. In medicine, treatment with anesthetics or electroconvulsive shock disrupts early memory but not long-term memory. Thus, anesthesia-sensitive memory (ASM) appeared to be consolidated into an anesthesia-resistant memory (ARM). We confirmed earlier reports of this phenomenon in *Drosophila* (Quinn and Dudai 1976; Tempel et al. 1983)[6,7] and turned our attention to the learning/memory mutants *dunce*, *rutabaga*, and *amnesiac*. The former two appeared to affect learning and memory within the first hour after training, whereas learning was near-normal in *amnesiac* and its maximum memory defect occurred later than those for *dunce* and *rutabaga* (Tully and Quinn 1985).[2] Importantly, it appeared as though ARM also was normal in the *amnesiac* mutant. Together, these observations suggested a genetic dissection of ASM from ARM and of ASM into two distinct temporal phases: short-term memory (STM) and middle-term memory (MTM) (Tully 1988; Tully et al. 1990).[8,9]

In addition to its anesthesia resistance, consolidated "long-term memory" (LTM) in other animals also appeared to depend on the synthesis of new proteins, although the experimental results were not consistent and this finding remained controversial (Davis and Squire 1984).[10] We were perplexed to discover, however, that ARM in flies was not disrupted by protein synthesis inhibitors (Boynton and Tully 1990).[11] Randolf Menzel in Berlin reported similar findings for olfactory memory in bees, leading us both to conclude that LTM in insects and mammals may derive from distinct mechanisms, a result quite contrary to our behavior-genetic expectations. Eric Kandel didn't believe it. He took me aside after delivering this message during a conference at Yale and said, "Tim, you simply haven't yet produced LTM in flies. It must be there; figure it out!" To do so, I dove into the old studies of memory. I was struck by one of the earliest by Hermann Ebbinghaus

(1885),[12] in which he showed that (his) memory consolidation required repeated practice sessions spaced over time. After several months of effort, Thomas Preat in my lab produced a crude inkling that extended training of *Drosophila* might yield a memory lasting more than one day. Extended training was so arduous, however, that any systematic study was beyond Preat's endurance. Thus, I began to design a "robotrainer" to study LTM in *Drosophila*.

Armed with Jim Watson's encouragement and financial support, robotrainer version II was built in a few months after my arrival at CSHL. Over the next 18 months, we showed in quick succession that (1) ten massed training sessions (no rest interval between training sessions) produced longer-lasting memory (days) than a single training session (less than a day), (2) ten spaced training sessions (15-min rest interval between each training session) produced an LTM lasting more than a week, (3) one-day memory after spaced training was disrupted by the protein synthesis inhibitor, cycloheximide, whereas one-day memory after massed training was not, and (4) ARM and one-day memory after massed training were abolished in the *radish* mutant, but a protein-synthesis-dependent LTM formed nonetheless (Tully et al. 1994).[13] These experiments established the behavioral (training) context required to induce a bona fide protein-synthesis-dependent LTM and extended our dissection of memory to show that ARM was genetically distinct from LTM, a biological subtlety never before realized.

Jerry Yin, then a postdoc in Chip Quinn's lab at MIT, was extremely excited to learn of our success in producing a protein-synthesis-dependent LTM in *Drosophila*. Recent biochemical and molecular work had shown that the *dunce* and *rutabaga* mutations disrupted two different components of the cAMP signaling pathway: cAMP-specific phosphodiesterase and adenylyl cyclase, respectively (Chen et al. 1986; Levin et al. 1992).[14,15] Jerry decided to clone the fly homolog of the cAMP response-element binding protein (CREB), the only transcription factor known to be activated by cAMP cell signaling. Jerry hypothesized that because CREB regulates the expression of other genes, it might be specifically involved in the formation of protein-synthesis-dependent LTM. Jerry's molecular and cellular characterizations of *dCREB2* revealed a gene that was alternatively spliced into (at least) two protein isoforms, one of which functioned as an activator of CRE-dependent transcription and the other of which functioned as a repressor of the activator (Yin et al. 1995).[16] In late 1992, he sent us transgenic flies carrying a CREB-repressor construct driven by the *hsp70*-inducible promoter (*hsp-CREBr*). With these flies, we could grow otherwise normal flies to adulthood (*dCREB2* is an essential gene) and then induce widespread expression of *hsp-CREBr* by exposing adults to a heat shock a few hours before training.

Jerry had spent several difficult years cloning and characterizing *dCREB2* and was running out of financial support from Quinn. By spring of 1993, I had been at CSH Laboratory for more than a year and still had not succeeded in getting any appreciable funding for my fly work. Again with Jim's support, however, we came up with the funds necessary to bring Jerry to CSHL to continue his work on CREB. Within weeks of arriving at CSHL, we first observed that induced expression of CREB-r specifically blocked LTM without affecting learning or ARM—a remarkable confirmation of his original hypothesis! Moreover, these results represented a genetic dissection reciprocal to the *radish* studies—disruption of CREB abolished LTM without affecting ARM. Thus, ARM and LTM represented functionally independent forms of memory.

Throughout this tense time, with funds waning, Jerry and I would stay late discussing various strategies to keep the work going. "Data speak!" emerged as the battle cry. When the results came in, we were giddy as schoolgirls over our expectation that the hard times were over.

I quickly called my friend and colleague, Jeff Hall, back at Brandeis to tell him what we had done. "My god, you've brought the psychology of memory to its biological knees!" he proclaimed. "But, tell me, did you do the experiments blind?" "Oh, Jeff," I blurted, "the experiments are done in the dark. Hundreds of flies are involved. Nobody can tell what's happening under such circumstances." "I understand," he replied, "but that's not the issue. This result is too important to leave open to obvious criticism. The issue is that you *must* be able to tell reviewers that the experiments were done blind." My heart sank. Shaking my head, I mumbled, "You're right. You're right, goddamnit. We have to repeat it all." And repeat it all we did. Over the course of another full year, every experiment eventually published in Tully et al. (1994)[13] and Yin et al. (1994)[17] was repeated with the experimenter(s) blind to genotype. And every result repeated.

Notes and References

1. Quinn W.G., Harris W.A., and Benzer S. 1974. Conditioned behavior in *Drosophila melanogaster*. *Proc. Natl. Acad. Sci.* **71**: 708–712.

2. Tully T. and Quinn W.G. 1985. Classical conditioning and retention in normal and mutant *Drosophila melanogaster*. *J. Comp. Physiol. [A]* **157**: 263–277.

3. Dubnau J. and Tully T. 1998. Gene discovery in *Drosophila*: New insights for learning and memory. *Annu. Rev. Neurosci.* **21**: 407–444.

4. Dubnau J., Chiang A.S., Grady L., Barditch J., Gossweiler S., McNeil J., Smith P., Buldoc F., Scott R., Certa U., et al. 2003. The staufen/pumilio pathway is involved in *Drosophila* long-term memory. *Curr. Biol.* **13**: 286–296.

5. Benzer S. 1973. Genetic dissection of behavior. *Sci. Am.* **229**: 24–37.

6. Quinn W.G. and Dudai Y. 1976. Memory phases in *Drosophila*. *Nature* **262**: 576–577.

7. Tempel B.L., Bonini N., Dawson D.R., and Quinn W.G. 1983. Reward learning in normal and mutant *Drosophila*. *Proc. Natl. Acad. Sci.* **80:** 1482–1486.

8. Tully T. 1988. On the road to a better understanding of learning and memory in *Drosophila*. In *Modulation of synaptic transmission and plasticity in the nervous system NATO ASI series, G* (ed. G. Hertting and H.-C. Spatz), pp. 401–417. Springer-Verlag, Berlin.

9. Tully T., Boynton S., Brandes C., Dura J.M., Mihalek R., Preat T., and Villella A. 1990. Genetic dissection of memory formation in *Drosophila melanogaster*. *Cold Spring Harbor Symp. Quant. Biol.* **55:** 203–211.

10. Davis H.P. and Squire L.R. 1984. Protein synthesis and memory: A review. *Psychol. Bull.* **96:** 518–559.

11. Boynton S. and Tully T. 1990. Induction of a long-term memory in *Drosophila* is not affected by protein synthesis inhibitors. In *Symposia Medica Hoechst 23: The biology of memory* (ed. E. Lindenlaub), pp. 91–102. Verlag, Berlin.

12. Ebbinghaus published his work in 1885: Ebbinghaus H. 1885. *Über das Gedchtnis. Untersuchungen zur experimentellen Psychologie*. Leipzig: Duncker & Humblot. The English edition is Ebbinghaus H. 1913. *Memory. A contribution to experimental psychology*. New York: Teachers College, Columbia University.

13. Tully T., Preat T., Boynton S.C., and Del Vecchio M. 1994. Genetic dissection of consolidated memory in *Drosophila*. *Cell* **79:** 35–47.

14. Chen C.N., Denome S., and Davis R.L. 1986. Molecular analysis of cDNA clones and the corresponding genomic coding sequences of the *Drosophila dunce+* gene, the structural gene for cAMP phosphodiesterase. *Proc. Natl. Acad. Sci.* **83:** 9313–9317.

15. Levin L.R., Han P.L., Hwang P.M., Feinstein P.G., Davis R.L., and Reed R.R. 1992. The *Drosophila* learning and memory gene rutabaga encodes a Ca^{2+}/Calmodulin-responsive adenylyl cyclase. *Cell* **68:** 479–489.

16. Yin J.C., Wallach J.S., Wilder E.L., Klingensmith J., Dang D., Perrimon N., Zhou H., Tully T., and Quinn W.G. 1995. A *Drosophila* CREB/CREM homolog encodes multiple isoforms, including a cyclic AMP-dependent protein kinase-responsive transcriptional activator and antagonist. *Mol. Cell Biol.* **15:** 5123–5130.

17. Yin J.C., Wallach J.S., Del Vecchio M., Wilder E.L., Zhou H., Quinn W.G., and Tully T. 1994. Induction of a dominant negative CREB transgene specifically blocks long-term memory in *Drosophila*. *Cell* **79:** 49–58.

Cell, Vol. 79, 49–58, October 7, 1994, Copyright © 1994 by Cell Press

Induction of a Dominant Negative CREB Transgene Specifically Blocks Long-Term Memory in Drosophila

J. C. P. Yin,* J. S. Wallach,† M. Del Vecchio,*
E. L. Wilder,‡ H. Zhou,* W. G. Quinn,†§ and T. Tully*
*Cold Spring Harbor Laboratory
Cold Spring Harbor, New York 11724
†Department of Brain and Cognitive Science
§Department of Biology
Massachusetts Institute of Technology
Cambridge, Massachusetts 02139
‡Department of Genetics
Howard Hughes Medical Institute
Harvard Medical School
Boston, Massachusetts 02115

Summary

Consolidated memory after olfactory learning in Drosophila consists of two components, a cycloheximide-sensitive, long-term memory (LTM) and a cycloheximide-insensitive, anesthesia-resistant memory (ARM). Using an inducible transgene that expresses a dominant negative member of the fly CREB family, LTM was specifically and completely blocked only after induction, while ARM and learning were unaffected. These results suggest that LTM formation requires de novo gene expression probably mediated by CREB family genes.

Introduction

A recurrent finding from work on the biology of learning and memory is the central involvement of the cAMP signal transduction pathway. In Aplysia, the cAMP second-messenger system is critically involved in neural events underlying both associative and nonassociative modulation of a behavioral reflex (Kandel and Schwartz, 1982; Kandel et al., 1987; Byrne et al., 1993). In Drosophila, two mutants (*dunce* and *rutabaga*) were isolated in a behavioral screen for defects in associative learning and are lesioned in genes directly involved in cAMP metabolism (Quinn et al., 1974; Dudai et al., 1976; Byers et al., 1981; Livingstone et al., 1984; Chen et al., 1986; Levin et al., 1992). These latter observations were extended with a reverse-genetic approach using inducible transgenes expressing peptide inhibitors of cAMP-dependent protein kinase A (PKA) and with analyses of mutants in the PKA catalytic subunit (Drain et al., 1991; Skoulakis et al., 1993). Recent work on mammalian long-term potentiation (LTP) also has indicated a role for cAMP in synaptic plasticity (Frey et al., 1993; Huang and Kandel, 1994; Bourtchuladze et al., 1994 [this issue of *Cell*]).

The formation of long-lasting memory in animals and of long-term facilitation in Aplysia can be disrupted by drugs that interfere with transcription or translation (Agranoff et al., 1966; Barondes and Cohen, 1968; Davis and Squire, 1984; Rosenzweig and Bennett, 1984; Montarolo et al., 1986). This suggests that memory consolidation requires

de novo gene expression. Considered along with the involvement of the cAMP second-messenger pathway, this requirement for newly synthesized gene products suggests a role for cAMP-dependent gene expression in long-term memory (LTM) formation.

In mammals, a subset of genes from the CREB/ATF family are known to mediate cAMP-responsive transcription (Habener, 1990; de Groot and Sassone-Corsi, 1993). cAMP-responsive element–binding proteins (CREBs) are members of the basic region–leucine zipper transcription factor superfamily (Landschulz et al., 1988). The leucine zipper domain mediates selective homodimer and heterodimer formation among family members (Hai et al., 1989; Hai and Curran, 1991). CREB dimers bind to a conserved enhancer element (CRE) found in the upstream control region of many cAMP-responsive mammalian genes (Yamamoto et al., 1988). Some CREBs become transcriptional activators when specifically phosphorylated by PKA (Gonzalez and Montminy, 1989; Foulkes et al., 1992), while others, isoforms from the CREM gene, are functional antagonists of these PKA-responsive activators (Foulkes et al., 1991; Foulkes and Sassone-Corsi, 1992).

Work in Aplysia has shown that cAMP-responsive transcription is involved in long-term synaptic plasticity (Schacher et al., 1988; Dash et al., 1990). A primary neuronal coculture system has been used to study facilitation of synaptic transmission between sensory and motor neurons comprising the monosynaptic component of the Aplysia gill-withdrawal reflex. Injection of oligonucleotides containing CRE sites into the nucleus of the sensory neuron specifically blocked long-term facilitation (Dash et al., 1990). This result suggests that titration of CREB activity might disrupt long-term synaptic plasticity.

We cloned a Drosophila CREB gene, *dCREB2*, to facilitate genetic manipulation of cAMP-responsive transcription in flies (Yin et al., submitted). This gene produces several isoforms that share overall structural homology and nearly complete amino acid identity in the basic region–leucine zipper with mammalian CREBs. The dCREB2-a isoform is a PKA-responsive transcriptional activator, whereas the dCREB2-b product blocks PKA-responsive transcription by dCREB2-a in cell culture (Yin et al., submitted). These molecules with opposing activities are similar in function to isoforms of the mammalian CREM gene (Foulkes et al., 1991, 1992; Foulkes and Sassone-Corsi, 1992). The numerous similarities in sequence and function between *dCREB2* and mammalian CREBs suggest that cAMP-responsive transcription is evolutionarily conserved.

To investigate the role of CREBs in LTM formation in Drosophila, dominant negative transgenic lines that express *dCREB2-b* under the control of a heat-shock promoter (*hs-dCREB2-b*) were generated. Groups of flies, which had been heat-shock induced or left uninduced, were tested for memory retention after Pavlovian olfactory learning. This acute induction regimen minimized potential complications from inappropriate expression of dCREB2-b

CANCER

The Road to Understanding the Origins of Cancer in Humans

Arnold J. Levine

T HE GREAT MAJORITY OF THE SCIENTISTS who were trained and worked during the 1960s through the 1980s, and who were interested in understanding the origins and nature of cancer, were the children of the molecular biology revolution and its practice at Cold Spring Harbor Laboratory. As a graduate student in the earlier part of the decade of the 1960s, I wanted to work in cancer biology, but all we knew for sure at that time was that cancers could be caused by some viruses, some chemicals, and, in a few rare examples, some types of genetic predisposition. This was a confusing set of facts and relationships, and in response to it, researchers worked only in one of these three different areas, largely in isolation from the other fields, publishing in different journals, and attending different meetings. There were few, if any, unifying ideas and little to indicate which of these areas was really important for understanding the origins of human cancer.

Clearly, the concepts that finally were to yield a satisfactory synthesis of these ideas came from all three fields, but the first and most

---

ARNIE LEVINE completed his Ph.D. at the University of Pennsylvania in 1966 and was a postdoctoral fellow at the California Institute of Technology, 1966–1968. After moving to Princeton in 1968, he began studying tumor viruses, in particular SV40, and in 1979, he and others discovered the p53 tumor suppressor protein. After consecutive appointments at the State University of New York, Stony Brook, Princeton University, and The Rockefeller University where he was President and Chief Executive Officer, Levine moved to the Institute for Advanced Study in 2002. A frequent visitor to Cold Spring Harbor Laboratory, Levine was a trustee of Laboratory from 1997 to 1998.

169

important insights came from the study of tumor viruses and developing systems in cell culture (cellular transformation) to mimic the events of tumor formation. This started with the retroviruses, utilizing Rous sarcoma virus (RSV) to produce tumors in chickens (1911) and transform cells in culture. In the search for other viruses that cause cancers, many other retroviruses of animals were isolated and a second class of viruses, the DNA tumor viruses, were found. The rules of molecular biology said that "work on the simplest model systems" utilizes the evolving new tools of molecular biology (cloning, sequencing, site-specific mutagenesis, etc.) and uses genetics to ask the right questions: How do these viruses transform cells? What viral genes are involved in this process? What are the functions of those genes? Are these observations meaningful in understanding cancers in humans?

The pathway to answer these questions began in the laboratory of Michael Bishop and Harold Varmus with a small piece of RNA (and its c-DNA clone) from the RSV that was present in transforming strains of RSV but not in the closely related parent virus that failed to transform cells. This indicated that the *SRC* gene could transform cells by itself without the other viral genes. Remarkably, a very close copy (with a few mutational changes in the viral or v-*SRC* gene) of the *SRC* gene was found in the DNA from chickens and even from human cells. These observations gave us the first reasonable synthesis for the roles of viruses (they can carry the SRC gene), chemicals (that cause the mutations), and genes (in the genome of chickens or humans) in the causation of cancers. It provided a self-consistent hypothesis to explain all the observations for the origins of cancers, even in humans.

As a graduate student at Columbia University, Michael Wigler, in the laboratories of Richard Axel and Saul Silverstein, developed and improved the techniques for getting foreign DNA into cells in culture. Robert Weinberg's group at the Massachusetts Institute of Technology employed this technique to introduce DNA from chemically transformed rodent cells into "normal cells" and showed that they could identify rare clones of cells that were transformed by this DNA, suggesting that this would be a good assay to isolate oncogenes like the one found in the *SRC* experiments. Wigler's group, now at Cold Spring Harbor, took on the challenge to transfect the DNA from cells derived from a human bladder carcinoma (T24) into mouse cells that behaved as nontransformed cells (NIH-3T3 cells). Employing novel strategies to dilute and isolate the human DNA in a mouse genetic background, and to clone and select for the DNA that transformed these cells, a transforming human gene was isolated that later proved to be one of the *RAS* oncogenes, the first oncogene shown to be associated with human cancers and cell transformation. Normal copies of this gene were shown to be present in the DNA from normal cells. Several human tumors gave rise to different oncogenes (based on restriction enzyme experiments), whereas three of the human tumors tested seemed to contain the same DNA element as this transforming principle (Goldfarb et al. 1982[1]; see also Goldfarb, p. 175). These results put in place the methods, ideas, and approaches that would be followed by many for the isolation of oncogenes that could well play a role in the origins of human cancers. (Today, these approaches have been replaced by high-volume DNA sequencing either of candidate genes or of a cancer genome.)

Wigler and his group went right on to the next question: What is the function of the normal and mutated (oncogene) *RAS* oncogene? To study this, they moved into a favorable (the simplest system) genetic system, yeast, under the presumption of the unity of biological functions over billions of years of evolution. The deletion of the two *Ras* genes from yeast was lethal and could be rescued by putting in a normal human *RAS* gene (c-DNA) or a *Ras* c-DNA oncogene from a tumor cell (Kataoka et al. 1985[2]; see also Powers, p. 187). Interestingly, there were differences in the behavior of the yeast with these two different genes. The oncogene failed to produce a cell cycle arrest after starvation of these cells, whereas the normal allele did arrest the cell cycle. After many years of research using yeast genetics, the downstream effectors of yeast *Ras* (cyclic-AMP-mediated) were shown to be different from those of human *RAS* (MAP kinases). But all was not lost from this exercise in great science; every one of the modifications of the Ras protein, from Ras activation to its insertion into the plasma membrane, were discovered in yeast and then confirmed in humans. It is these protein modifications that remain our best approach to finding drugs that inhibit the *RAS* oncogene.

From this body of research, in which DNA from tumors was employed to transform cells in culture, two classes of oncogenes were commonly isolated: those located at the plasma membrane of the cell (receptors, Ras G proteins) and those in the cell nucleus (Myc), which were eventually shown to be transcription factors. The search for transcription factors that bound to specific DNA sequences and enhanced the rate of transcription of a gene got a good start at Cold Spring Harbor with the early work of Robert Tjian (first with SV40 T antigen and SV40 DNA [see Tjian, *Tumor Viruses*, p. 119]). Tjian found that the SV40 genome contained a DNA sequence that was recognized by the HeLa cell (one could grow lots of HeLa cells to make extracts of these proteins) activator protein-1 (AP-1). The search for cellular proteins that bound to this AP-1 site led to two oncogenes, *Fos* and *Jun*, which had been described previously in retroviruses. These studies pioneered new technologies (DNA affinity precipitation) that permitted one to dissect the many protein forms of an oncogene such as *Fos* using two-dimensional gel electrophoresis (Franza et al. 1988[3]; see also Franza, p. 199).

Whereas the retroviruses contained transforming oncogenes that were in each case derived from the cellular genome, the DNA tumor viruses contained oncogenes that could transform cells in culture, but they were not recently derived from the cellular genome (i.e., they had no DNA homology with cellular genes). Unlike the oncogenes from retroviruses, these viral genes were essential for the replication of the DNA viruses. Using viral genetics, a number of groups demonstrated that the SV40 large T antigen, the adenovirus E1A and E1B proteins, and the papillomavirus E6 and E7 proteins were required by these viruses to transform cells in culture. How do these genes transform cells in culture and form tumors in animals? The answers to these questions have their origins in the 1970s, when somatic cell fusions between normal cells and cancer cells and a possible explanation of how retinoblastomas arise in children came together to suggest that wild-type alleles of certain genes in the human genome can prevent cancers or the transformation of cells in culture. If both copies of these tumor suppressor alleles were lost to mutation, then the probability of developing a cancer would increase. By the late 1970s and into the 1980s, candidate tumor suppressors—the retinoblastoma protein (Rb) and the p53 protein—had been identified and their genes cloned. The demonstration that the adenovirus E1A protein binds specifically to the Rb protein, by the Harlow group at Cold Spring Harbor (Whyte et al. 1988[4]; see also Harlow, p. 203), was particularly important for our understanding of how the DNA viral oncogene products bind to and inactivate a tumor suppressor protein. At the time, the SV40 T antigen–p53 and adenovirus E1B–p53 protein complexes were known to exist, but p53 was not yet thought to be a tumor suppressor (that was demonstrated one year later in 1989) so the Harlow paper introduced a novel concept: In human tumors, the *Rb* gene was inactivated by mutations, but in tumors originating from the DNA tumor viruses, *Rb* was inactivated in a protein complex. These observations help to explain how human papillomaviruses can cause cancers in men and women. On the basis of this work, it was becoming clear that mutations in multiple oncogenes and tumor suppressor genes were required to originate cancers in humans. This concept was predicted by an earlier (1983) demonstration that at least two different oncogenes (the adenovirus *E1A* and *RAS*) were required to fully transform primary rat embryo fibroblasts in culture (Ruley 1983[5]; see also Ruley, p. 181). Each of these gene products contributes a different activity: E1A inactivates Rb and immortalizes cells for growth in culture and Ras activates the MAP kinase pathway and sends the cell into a division cycle while providing a transformed cell phenotype of foci formation, growth in agar, and the potential to form tumors in animals. For primary human cells in culture, transformation is complex, requiring SV40 large T antigen (inactivates *Rb* and *p53*), small t antigen (acts upon the PP2A phosphatase), Ras (activates MAP kinases and Ets and AP-1 transcription factors), and telomerase (repairs the DNA ends of chromosomes).

This brief review of a selected portion of cancer research from 1970 to 1990 shows that Jim Watson clearly knew how to pick research topics that were important, that were ready for exploration, and that would lead to new avenues of understanding and an extension into human cancer biology. In 1968, when Watson came to Cold Spring Harbor, he identified the DNA tumor viruses as a focus for research (see Sambrook, *Tumor Viruses*, p. 89). He assembled a great group of young scientists (not a bunch of professors) to lead this effort and it thrived, drawing into the field some of the best young biologists from all over the world and opening up new fundamental mechanisms (such as splicing, cell cycle regulation, and transcriptional gene regulation in higher eukaryotes), as well as a new understanding of cancer biology.

I first met Jim in 1967, the year before he went to Cold Spring Harbor, in Aspen, Colorado, where he gave the first lecture in a course on the regulation of gene expression at the transcriptional level in *Escherichia coli* and its phages (σ factors and RNA polymerase). It was clear from his talk that a few years before, at Harvard, he had picked the right next topic (how genes are regulated) for exploration in his laboratory. By the summer of 1969, I attended the first Tumor Virus Course at Cold Spring Harbor (as a student). Later that summer, I gave a talk in the first Tumor Virus Meeting at the Lab. Over the years, I have been a teacher in those courses, a trustee of the laboratory, a collaborator in science, and a good friend to one of the special places that forms a cornerstone of the cancer research community.

Contributing new concepts to human cancer biology and opening paths to new treatments for cancer were always high on the list of the Laboratory's goals. The development of new drugs to treat cancers suffered from the fact that tractable simple systems for drug screening, such as transformed cells in culture, or cells derived from cancers, behaved differently in response to a drug treatment than did a living human being with cancer. What was needed was a major effort to discover animal models of cancers that would more faithfully resemble human cancers, yield new information, and provide a means to test new drugs. One of the first to move into this space was Doug Hanahan. When the SV40 genome was first placed into one of the chromosomes of a transgenic mouse, it produced papillomas of the choroid plexes largely because the viral enhancer sequences for transcription appear to function best in that tissue. Hanahan employed a new tissue-specific enhancer for pancreatic β cells regulating SV40 large T antigen (inactivating p53 and Rb) to initiate tumors of the pancreas (the first neuroendocrine tumors of mice) (Hanahan 1985[6]; see also Hanahan, p. 193). This experimental system would go on to be one of the best ways to study those factors that regulate tumor angiogenesis and to test drugs and conditions to block the invasion of blood vessels in a model tumor.

To this day, the Laboratory at Cold Spring Harbor continues to contribute important new concepts and animal models to cancer research. The early labs of the 1970s have largely been turned over to a new generation of young scientists, while the postdoctoral fellows trained at Cold Spring Harbor in the 1970s and

1980s populate professorships at the great universities all over the world. Most important, we now know what causes cancers in human beings (this is the 40th anniversary of that effort) and drugs are being designed to target the abnormal gene products of oncogenes, reactivate dormant tumor suppressor gene products, and even utilize the adenoviruses as vectors for gene therapy to introduce normal tumor suppressor genes into human cancers. This is a triumph of science and modern medicine and a triumph of the people who populated the laboratories all over the world and made this happen.

Notes and References

1. Goldfarb M., Shimizu K., Perucho M., and Wigler M. 1982. Isolation and preliminary characterization of a human transforming gene from T-24 bladder carcinoma cells. *Nature* **296:** 404–409.

2. Kataoka T., Powers S., Cameron S., Fasano O., Goldfarb M., Broach J., and Wigler M. 1985. Functional homology of mammalian and yeast RAS genes. *Cell* **40:** 19–26.

3. Franza B.R. Jr., Rauscher F.J. 3rd, Josephs S.F., and Curran T. 1988. The Fos complex and Fos-related antigens recognize sequence elements that contain AP-1 binding sites. *Science* **239:** 1150–1153.

4. Whyte P., Buchkovich K.J., Horowitz J.M., Friend S.H., Raybuck M., Weinberg R.A., and Harlow E. 1988. Association between an oncogene and an anti-oncogene: The adenovirus E1A proteins bind to the retinoblastoma gene product. *Nature* **334:** 124–129.

5. Ruley H.E. 1983. Adenovirus early region 1A enables viral and cellular transforming genes to transform primary cells in culture. *Nature* **304:** 602–606.

6. Hanahan D. 1985. Heritable formation of pancreatic beta-cell tumours in transgenic mice expressing recombinant insulin/simian virus 40 oncogenes. *Nature* **315:** 115–122.

MITCHELL GOLDFARB

Goldfarb M., Shimizu K., Perucho M., and Wigler M. 1982. **Isolation and preliminary characterization of a human transforming gene from T24 bladder carcinoma cells.** *Nature* **296**: 404–409. (Reprinted, with permission, from Macmillan Publishers Ltd ©1982.)

T HE FIRST DISCOVERIES OF HUMAN CANCER GENES were made in the early 1980s in several pioneering labs, including that of Michael Wigler[1] at Cold Spring Harbor. The paper reported the first molecular cloning of an activated oncogene from a human tumor cell line. I'd like to offer some scientific and personal perspective to the events leading to this discovery and to events shortly thereafter.

Three findings during the period of 1975–1980 set the stage for the discovery of human oncogene(s). First, the labs of Michael Bishop and Harold Varmus showed that the human genome harbors so-called proto-oncogenes.[2] The key experimental discovery was of a cellular gene (c-src) that is a very close relative of an oncogene (v-src) which had been known for years to be part of the genome of a retrovirus. v-src differs from c-src by a carboxy-terminal coding sequence truncation and by association with a robust retroviral promoter that drives expression of the v-src gene to high levels. With this finding, it was reasonable to speculate that cancers of nonviral etiology could result from mutations that convert proto-oncogenes to oncogenes with altered regulation or protein functions.

Second, Robert Weinberg's lab at the Massachusetts Institute of Technology showed that chemically transformed rodent cells contain activated cellular oncogenes, as demonstrated by induction of transformed foci after transfection of cultured fibroblasts with DNA isolated from the chemical transformants. This

MITCH GOLDFARB came from Bob Weinberg's laboratory at MIT to work with Mike Wigler at Cold Spring Harbor as a postdoctoral fellow. This article describes his pivotal role in isolating activated *ras* oncogenes from human cancer cells. Mitch stayed for about 4 years at Cold Spring Harbor, rising progressively through the ranks to the level of Senior Staff Investigator. In 1984, Mitch left the Laboratory for a position at the College of Physicians & Surgeons of Columbia University. Later a professor at the Brookdale Center for Molecular Biology of Mt. Sinai School of Medicine, he is now Professor of Biology in the Department of Biological Sciences at Hunter College. goldfarb@genectr.hunter.cuny.edu

discovery grew out of my thesis project as a graduate student in Bob's lab. I had been studying the ability of the DNA of a replication-defective transforming retrovirus (Harvey sarcoma virus) to integrate into the cellular genome.[3] When the double-stranded HaSV DNA replication intermediate was purified from infected cells and transfected into fibroblasts, it integrated and induced morphological transformation of the cells. Early on, we reasoned (incorrectly) that I had recapitulated the precise mechanism by which the retroviral DNA integrates in infected cells. In fact, this turned out to be an early demonstration of random DNA breakage and integration that are characteristic of naked DNA transfection. More interestingly, however, genomic DNA isolated from the primary transformants could be used to generate secondary transformants, and the process repeated yet again to generate tertiary transformants. Since the secondary transformants contained only a single integrated HaSV genome, transfection was clearly efficient enough to achieve transfer of a biologically active, single-copy oncogene to a recipient cell. Bob suggested that such efficiency made feasible searches for cellular oncogenes activated in cancer or chemically transformed cells. I was skeptical, since we still harbored the belief that proviral DNA had specialized features that facilitated integration upon transfection. Nevertheless, I undertook the "side project" of transfecting DNA from chemically transformed fibroblast lines, and after screening a couple of dozen, I found a line with transforming capacity. The project was then continued full time by Chiaho Shih and Benny Shilo. The resultant publication[4] made feasible future searches for human oncogenes in cancers.

Third, Michael Wigler was investigating the phenomenon of cellular gene transfection, initially in the labs of Richard Axel and Saul Silverstein at Columbia University and later in his own new lab at Cold Spring Harbor. Using a selective medium (HAT medium) to detect transfer of the thymidine kinase (TK) gene, Mike showed that transfection of total cell DNA resulted in stable transfer of TK to recipient cells.[5] Furthermore, recipient cells receiving the selected TK also received other random unselected segments of cotransferred DNA, with the TK and cotransferred DNA elements linked to one another before integration into the host genome. These findings had many technical ramifications, including the ability to molecularly tag a target gene by including generous amounts of a tag element in DNA transfection mixtures.[6] The next step was to apply these methods to the tagging and cloned retrieval of cellular oncogenes.

Upon my arrival as a postdoc in Mike Wigler's lab, another postdoc, Manuel Perucho,[7] and I got busy with transfections, screening DNAs from a large bank of human tumor cell lines previously assembled by Jorgen Fogh.[8] In these screens, approximately 25% of the tumor lines harbored selectable oncogenes that gave transformed foci on transfected 3T3 fibroblasts. Manuel used genetic fingerprinting to determine whether different tumor lines yielded different or identical onco-

genes.[9] I focused on cloning the oncogene from the T24 line of bladder carcinoma, since this gene transferred at very
high efficiency and was small enough to
be cloned in a single bacteriophage vector. By this time, a third postdoc, Kenji
Shimizu,[10] had perfected the technique of
suppressor rescue, whereby a bacteriophage cloning vector carrying an amber
mutant could grow on a suppressor-free
Escherichia coli host only if a cloned DNA
segment itself contained an amber suppressor (*supF*). By mixing T24 DNA with
supF DNA prior to transfection onto 3T3
fibroblasts, I obtained primary transformants with multiple *supF* elements, and
some secondary and tertiary transformants carried a single linked *supF*. Sup-

Mitch Goldfarb, 1983. *(Photograph by Herb Parsons, courtesy of Cold Spring Harbor Laboratory Archives.)*

pressor rescue simplified the cloning of the T24 oncogene, as described in the paper.

Our paper was rapidly followed by many similar studies from several labs. Almost all oncogenes detected by transfection were shown to be members of the *ras* family that had been activated by mutation. Ironically, the activated oncogene in T24 cells was c-Ha-*ras*, the cell homolog of v-Ha-*ras* responsible for oncogenicity of HaSV. And as a bizarre twist of circumstance, the EJ bladder carcinoma used in a parallel cloning effort from Bob Weinberg's group appears to have been contam-inated by T24 carcinoma cells in a prior lab that maintained both EJ and T24 cells. (Bob's source of EJ cells contained the same oncogene mutation found in T24, but EJ cells from other sources had no detectable activated oncogene.) So my early attempts to focus on cancer cells other than those studied by my former mentor were thwarted in a most unexpected way.

Notes and References

1. Mike Wigler was appointed to head the Mammalian Cell Genetics Group at Cold Spring Harbor Laboratory in late 1978. He obtained his Ph.D. from the College of Physicians & Surgeons of Columbia University, where he also carried out postdoc-toral research.

2. Spector D.H., Varmus H.E., and Bishop J.M. 1978. Nucleotide sequences related to the transforming gene of avian sarcoma virus are present in the DNA of uninfected vertebrates. *Proc. Natl. Acad. Sci.* **75:** 4102–4106.

3. Goldfarb M.P. and Weinberg R.A. 1981. Generation of novel, biologically active Harvey sarcoma viruses via apparent illegitimate recombination. *J. Virol.* **38:** 136–150.

4. Shih C., Shilo B.Z., Goldfarb M.P., Dannenberg A., and Weinberg R.A. 1979. Passage of phenotypes of chemically transformed cells via transfection of DNA and chromatin. *Proc. Natl. Acad. Sci.* **76:** 5714–5718.

5. Pellicer A., Wigler M., Axel R., and Silverstein S. 1978. The transfer and stable integration of the HSV thymidine kinase gene into mouse cells. *Cell* **14:** 133–141.

6. Wigler M., Sweet R., Sim G.K., Wold B., Pellicer A., Lacy E., Maniatis T., Silverstein S., and Axel R. 1979. Transformation of mammalian cells with genes from procaryotes and eukaryotes. *Cell* **16:** 777–785.

7. Manuel Perucho earned his Ph.D. in biological sciences at the University of Madrid, Spain in 1976. He did postdoctoral work at the Max-Planck Institut für Molekulare Genetik, Berlin, and then worked as a postdoctoral fellow in the Wigler laboratory between 1979 and 1982, when he left to take a position in the Microbiology Department at the Stony Brook campus of the State University of New York.

8. In the 1970s, Jorgen Fogh, while working at the National Cancer Institute, Bethesda, developed from human cancers well over 100 cell lines that could cause tumors in nude mice. These cell lines were inherited by the Wigler lab in the early 1980s.

9. Perucho M., Goldfarb M., Shimizu K., Lama C., Fogh J., and Wigler M. 1981. Human-tumor-derived cell lines contain common and different transforming genes. *Cell* **27:** 467–476.

10. Natalie Angier in her book *Natural Obsessions: Striving to Unlock the Deepest Secrets of the Cancer Cell* quotes Carmen Birchmeier, a scientist working in Mike Wigler's group, as saying "Mike (Wigler) was very lucky to have Kenji (Shimizu) around. Kenji was probably one of the only scientists alive who could have cloned the gene using that bacterial tag technique. Anybody else would have been crazy to try it." Carmen Birchmeier perhaps overstates the impact of the method, although its elegance is beyond dispute. Its main value was that it obviated the need for filter hybridization to screen plated libraries for nonselectable tags. The more fundamental finding by Mike Wigler's group was the ability to place *any* tag near a selected gene (oncogene) through the transfection process. Kenji Shimuzu worked at Cold Spring Harbor Laboratory from 1980 to 1983 and then retrurned to Japan, first to Kyushu University and most recently to the Department of Molecular Genetics at Okayama University.

eventually cover. In the absolute reference frame the ridges must move relative to their immediately underlying asthenospheric sources[19].

What is not clear is why the past positions of the ridges in the approximate absolute reference frame provided by South America seem to match the areas of uplift so well. One possibility is that because asthenospheric sources are probably vertically extensive[38,39] and have horizontal velocities much less than the overlying plates[38-40], the asthenospheric zone extending from vertically above the deep source of a new ridge to the present position of the ridge may all have become hotter than it was before ridge creation. If so, this increased temperature could provide the heat needed for the phase change.

In detail, this model must be modified. For example, the uplift of southern Africa has been treated as a uniform vertical uplift simultaneously affecting all of high Africa, yet it is quite clear from Fig. 1c, d that we would expect a wave of uplift to have moved across high Africa starting in the east, and taking a geologically significant time to reach the west. This problem will not be discussed here in detail because part of its solution depends on considering still earlier phases of uplift probably resulting from the breakup of Gondwanaland. Nevertheless, it is worth noting that there is a distinct structural and topographic

asymmetry to southern Africa. The Cenozoic Kalahari basin lies in the western half of southern Africa, while the eastern half consists primarily of Precambrian rocks: during much of Cenozoic time, the eastern half of southern Africa has been structurally raised relative to the western half. In addition, the topography is markedly asymmetric: the highest parts of southern Africa are in Lesotho, very close to the eastern margin. Here, and nowhere else, the topography rises to well over 3 km. This structural and topographic asymmetry is what would be expected from a more detailed application of the model. The same mechanism is envisaged for the uplift of south-east Australia, where the expected topographic asymmetry is quite clear.

All these speculations can be tested by better data from the uplift of land surfaces, high-pressure petrology and three-dimensional mantle flow models.

I thank P. England for providing a heat calculation programme and T. Holland for working out the likely values of the heats of reaction involved, also M. J. Bickle, E. Leitch, L. Nicolaysen, S. Richardson, F. J. Vine and N. H. Woodcock for comments and criticisms. Adele Prowse drew the figures. The production of the diagrams was greatly aided by the map-making system developed under a grant from the NERC GR3/2277. Cambridge Earth Sciences contribution no. 253.

Received 29 April 1981; accepted 27 January 1982.

1. Crough, S. T. *Geology* **9**, 2–6 (1981).
2. Parsons, B. & Sclater, J. G. *J. geophys. Res.* **82**, 803–827 (1977).
3. Kennedy, G. C. *Am. Scient.* **47**, 491 (1959).
4. O'Connell, R. J. & Wasserburg, G. J. *Rev. Geophys.* **5**, 329–410 (1967).
5. Gass, I. G., Chapman, D. S., Pollack, H. N. & Thorpe, R. S. *Phil. Trans. R. Soc.* **A288**, 581–597 (1978).
6. King, L. C. *The Morphology of the Earth* (Oliver & Boyd, Edinburgh, 1962).
7. Brown, C. & Girdler, R. W. *J. geophys. Res.* **85**, 6443–6455 (1980).
8. Fairhead, J. D. & Reeves, C. V. *Earth planet. Sci. Lett.* **36**, 63–76 (1977).
9. Sclater, J. G., Jaupart, C. & Galson, D. *Rev. Geophys. Space Phys.* **18**, 269–311 (1980).
10. Mercier, J.-C. C. *J. geophys. Res.* **85**, 6293–6303 (1980).
11. *The World, Mercator Projection*, Ser. 1142 (Department of Defense, Washington DC, 1972).
12. Stacey, F. D. *Physics of the Earth* 2nd edn (Wiley, New York, 1977).
13. Newton, R. C., Charlu, T. V. & Kleppa, O. J. *Geochim. cosmochim. Acta* **44**, 933 (1980).
14. Wood, B. J., Holland, T. J. B., Newton, R. C. & Kleppa, O. J. *Geochim. cosmochim. Acta* **44**, 1363 (1980).
15. Crough, S. T., Morgan, W. J., & Hargraves, R. B. *Earth planet. Sci. Lett.* **50**, 260–274 (1980).
16. McGetchin, T. R. & Silver, L. T. *J. geophys. Res.* **77**, 7022–7037 (1972).
17. Ferguson, J., Arculus, R. J. & Joyce, J. *BMR J. Austr. Geol. Geophys.* **4**, 227–241 (1979).
18. Duncan, R. A. *Tectonophysics* **74**, 29–42 (1981).
19. Solomon, S. C., Sleep, N. H. & Richardson, R. M. *Geophys. J. R. astr. Soc.* **42**, 769–801 (1975).
20. McElhinny, M. W. *Nature* **241**, 523–524 (1973).
21. Jurdy, D. M. & Van der Voo, R. *J. geophys. Res.* **79**, 2945–2952 (1974).
22. Briden, J. C., Hurley, A. M. & Smith, A. G. *J. geophys. Res.* **86**, 11631–11656 (1981).
23. Hager, B. H. & O'Connell, R. J. *J. geophys. Res.* **86**, 4843–4867 (1981).
24. Harper, J. F., Hurley, A. M. & Smith, A. G. *Tectonophysics* **74**, 169–187 (1981).
25. Harper, J. F. *Geophys. J. R. astr. Soc.* **55**, 87–110 (1978).
26. Weissel, J. K. & Hayes, D. E. *Earth planet. Sci. Lett.* **36**, 77–84 (1977).
27. Shaw, R. D. *Bull. Austr. Soc. exp. Geophys.* **9**, 75–81 (1978).
28. Andrews, E. C. *J. Proc. R. Soc. N.S.W.* **44**, 420–480 (1911).
29. King, L. C. *Trans. Proc. geol. Soc. S. Afr.* **62**, 113–138 (1959).
30. Browne, W. R. *J. geol. Soc. Austr.* **16**, 559–569 (1969).
31. Hills, E. S. *The Physiography of Victoria* (Whitcombe & Tombs, Melbourne, 1975).
32. Craft, F. A. *Proc. Linn. Soc. N.S.W.* **58**, 437–460 (1933).
33. Wellman, P. *J. geol. Soc. Austr.* **26**, 1–9 (1979).
34. Wellman, P. & McDougall, I. *Tectonophysics* **23**, 49–65 (1974).
35. Sutherland, F. L. *Tectonophysics* **48**, 413–427 (1978).
36. Weissel, J. K. & Watts, A. B. *J. geophys. Res.* **84**, 4572–4582 (1979).
37. Finlayson, D. M., Prodehl, C. & Collins, C. D. N. *BMR Austr. Geol. Geophys.* **4**, 243–252 (1979).
38. Hager, B. H. & O'Connell, R. J. *Tectonophysics* **50**, 111–133 (1978).
39. Hager, B. H., & O'Connell, R. J. *J. geophys. Res.* **84**, 1031–1048 (1979).
40. Parmentier, E. M. & Oliver, J. E. *Geophys. J. R. astr. Soc.* **57**, 1–21 (1979).

Isolation and preliminary characterization of a human transforming gene from T24 bladder carcinoma cells

Mitchell Goldfarb, Kenji Shimizu, Manuel Perucho & Michael Wigler

Cold Spring Harbor Laboratory, Cold Spring Harbor, New York 11724, USA

DNA from T24, a cell line derived from a human bladder carcinoma, can induce the morphological transformation of NIH 3T3 cells. Using techniques of gene rescue to clone the gene responsible for this transformation, we have found that it is human in origin, <5 kilobase pairs in size and is homologous to a 1,100-base polyadenylated RNA species found in T24 and HeLa cells. Blot analysis indicates extensive restriction endonuclease polymorphism near this gene in human DNAs.

THE progression of a cell lineage from normalcy to malignancy may involve the mutation or activation of one or more genes. The genomes of retroviruses contain candidates for such 'oncogenes'. Certain retroviruses capable of inducing neoplasia *in vivo* and cell transformation *in vitro* contain transduced cellular genes which entirely encode the oncogenic proteins of these viruses[1,2]. If these or other oncogenes are expressed in tumours of viral or nonviral origin, the introduction of these

genes into cultured cells might transform the recipients and render them tumorigenic. Indeed, DNA from some chemically transformed mouse cells can morphologically transform NIH 3T3 mouse fibroblasts following DNA-mediated gene transfer[3]. More recently, it has been reported that DNA from certain human tumour cell lines can also morphologically transform NIH 3T3 cells[4,5]. We have detected transforming activity in DNA from 5 of 21 human tumour cell lines[6]; the resulting

0028–0836/82/130404—06$01.00

Ruley H.E. 1983. **Adenovirus early region 1A enables viral and cellular transforming genes to transform primary cells in culture.** *Nature* **304:** 602–606. (Reprinted, with permission, from Macmillan Publishers Ltd ©1983.)

J OE SAMBROOK INVITED ME TO WORK AT COLD SPRING HARBOR during one of his visits to the Imperial Cancer Research Fund, where I was a postdoc in Mike Fried's laboratory.[1] My arrival at Cold Spring Harbor was fairly typical. Joe asks, "What are you doing here so soon?" I answer, and he then replies, "Don't be absurd, no one ever comes when they say they are going to." But with four feet of bench space in James Laboratory (a luxury in those days), outstanding colleagues, and the freedom to work on almost anything, I had all I could want.

I was initially interested in the idea that proteins and possibly carbohydrates on the cell surface regulated cellular behavior, for example, during embryonic development and following malignant transformation. So I set out to clone genes that might contribute to the relevant cell-cell interactions and things were going very well; however, after a few months, I was offered a promotion to Staff Investigator with a significant increase in salary, but with the stipulation that I work on a problem relevant to an NIH program project grant on DNA tumor viruses. The grant was a major source of funding but was short on personnel. I was no stranger to tumor viruses, having worked on mouse polyomavirus in London, but wanted to identify a significant problem before switching. The ideas did not take long to germinate.

By the end of the 1970s, DNA tumor viruses were known to encode genes capable of transforming cultured cells to a malignant state. In the case of the human ade-

---

EARL RULEY is a Professor of Microbiology and Immunology at Vanderbilt Medical Center, Nashville. During his entire career—from his postdoctoral days at the Imperial Cancer Research Fund in the late 1970s to faculty positions at Cold Spring Harbor and the Massachusetts Institute of Technology in the 1980s and then Vanderbilt—Earl has used a variety of genetic and biochemical approaches to study mechanisms important in growth control and cancer. In the 1984 Cold Spring Harbor Annual Report, Jim Watson described the work in this paper as "a major coup." earl.ruley@Vanderbilt.edu

Earl Ruley, 1983. *(Courtesy of Cold Spring Harbor Laboratory Archives.)*

noviruses types 2 and 5, a combination of two genes E1A and E1B together was said to result in "complete" transformation, whereas E1A alone resulted in "partial" transformation.[2] These transforming activities were assessed in primary cultures of baby rat kidney (BRK) cells, which like most primary cells (i.e., cells recently explanted from animals) had a limited life span in culture. Partial transformation by E1A enhanced the ability of primary BRK cells to proliferate in culture and facilitated the establishment of permanent cell lines. However, despite their robust growth in culture, the cells were not fully transformed because they did not form tumors in animals. The situation with human adenoviruses was reminiscent of then unpublished studies on mouse polyomavirus. Although the polyomavirus middle T antigen could transform established cell lines to a tumorigenic state, a combination of the small, middle, and large T antigens was required to transform primary embryonic fibroblasts.[3,4] Moreover, the polyomavirus large T antigen, like adenovius E1A, promoted the conversion of primary cells into established, but nontumorigenic, cell lines.[5] It therefore appeared that transformation of primary cells by adenoviruses and polyomavirus required a minimum of two viral proteins, each acting on different aspects of cellular growth control. One protein was responsible for immortalization of cells (i.e., for survival, self self-renewal, and responses to growth factors); a second protein was essential for development of malignancy. The concept of multistep transformation was also potentially important to the mechanism of transformation by human oncogenes. Several groups, including Mike Wigler's at Cold Spring Harbor, had isolated from human tumors mutated Ha-*RAS* oncogenes that transformed a permanent line of NIH-3T3 cells to a malignant state.[6] But it seemed implausible that something as simple as a single point mutation in a gene such as Ha-*RAS* could explain the biological complexity of tumor onset and progression in humans. Mutated Ha-*RAS* could efficiently transform established lines of cultured cells but, like the adenovirus E1B and polyomavirus middle T antigen, it might lack the functions required to transform primary cells. If so, perhaps a combination of mutated Ha-

RAS and a viral protein with an establishment function would together be sufficient to transform cultures of primary cells.

These predictions were straightforward and easily tested, and the results were published in the paper included in this volume. However, I also encountered several reasons not to do the experiments. Bill Topp[7] said the studies were unnecessary because in 6 months, his research would have solved the problem of oncogenic transformation. Mike Matthews[8] strongly cautioned against any experiment not involving biochemistry. Mike Wigler did not say anything but generously provided clones of the normal and mutated Ha-*RAS* genes. The manuscript was submitted to *Nature* just before the DNA Tumor Virus meeting in Cambridge, England.[9] Soon after arriving at the meeting, I went to London to revise the page proofs, and the paper was published prior to my return to the States. The rush allowed the paper to be published in the same issue as complementary studies by Hartmut Land in Bob Weinberg's laboratory[10] and by Robert Newbold and Bob Overell.[11] Subsequent studies[12–14] identified cellular targets of the viral oncogenes, notably the proteins encoded by the *RB, TP53,* and *p16INK* tumor suppressor genes, and also established the importance of oncogene-induced proliferation arrest (senescence) and programmed cell death (apoptosis) in multistep transformation. The small DNA tumor viruses rely on host machinery for DNA replication, and the inactivation of *RB* by genes such as E1A frees cells from the shackles of the G_1 phase of the growth cycle and drives them into S phase. This activates p53-dependent apoptosis, a mechanism to eliminate aberrantly cycling cells. The viruses encode proteins such as p55 (encoded by E1B) that bind to the p53 protein and prevent apoptosis, thus preventing the untimely death of the infected cell.

My time at Cold Spring Harbor influenced work in my laboratory for more than a decade. We showed that E1A enhances the stability of p53, thus inducing apoptosis.[15,16] However, E1B blocks p53-dependent apoptosis and transforms cells when combined with E1A. This suggested that p53 can function directly as a tumor suppressor by blocking the tumorigenic activities of an oncogene (E1A) rather than by acting indirectly as a guardian of the genome. And the story continued to evolve. Cells transformed by E1A and Ha-*RAS* express high levels of p53 and form tumors but remain susceptible to p53-dependent apoptosis. This leaves the cells highly sensitive to therapeutic agents used to treat cancer.[17] Thus, like apoptosis and malignancy, the susceptibility of tumor cells to irradiation and malignancy is affected by the interplay between oncogenes and tumor suppressor genes. These days, it may seem a commonplace that embraces between proteins should have such far-reaching consequences. Twenty-five years ago, the black box surrounding cancer was considerably larger and such insights were a source of wonderment.

Notes and References

1. In the 1970s and early 1980s, the Imperial Cancer Research Laboratories in Lincoln's Inn Fields, London, was the scientific home of a group working on polyoma, a virus whose size, genomic organization, and biological characteristics are similar to those of SV40. Because of this common interest, visits—particularly between scientists in Mike Fried's and Joe Sambrook's laboratories—were frequent.

2. Houweling A., van den Elsen P.J., and van der Eb A.J. 1980. Partial transformation of primary rat cells by the leftmost 4.5% fragment of adenovirus 5 DNA. *Virology* **105:** 537–550.

3. Treisman R., Novak U., Favaloro J., and Kamen R. 1981. Transformation of rat cells by an altered polyomavirus genome expressing only middle-T protein. *Nature* **292:** 595–600.

4. Rassoulzadegan M., Cowie A., Carr A., Glaichenhaus N., Kamen R., and Cuzin F. 1982. The roles of individual polyomavirus early proteins in oncogenic transformation. *Nature* **300:** 713–718.

5. Rassoulzadegan M., Naghashfar Z., Cowie A., Carr A., Grisoni M., Kamen R., and Cuzin F. 1983. Expression of the large T protein of polyomavirus promotes the establishment in culture of "normal" rodent fibroblast cell lines. *Proc. Natl. Acad. Sci.* **80:** 4354–4358.

6. See Goldfarb, p. 175, this volume.

7. Bill Topp (see footnote 7 in Botchan, p. 116).

8. Mike Matthews, an English biochemist, came to Cold Spring Harbor from the University of California at San Francisco in 1974. After working for 2 years in James Laboratory, Mike moved to larger quarters in Demerec Laboratory. He left Cold Spring Harbor in 1996 to take a position at the University of Medicine and Dentistry, New Jersey.

9. The DNA Tumor Virus meetings, which began in 1969, were held annually at Cold Spring Harbor until the early 1980s. From then on, the venue of the meetings alternated between Cambridge, U.K. and Cold Spring Harbor. When the field was at its peak, the meetings attracted several hundred participants.

10. Land H., Parada L.F., and Weinberg R.A. 1983. Tumorigenic conversion of primary embryo fibroblasts requires at least two cooperating oncogenes. *Nature* **304:** 596–602.

11. Newbold R.F. and Overell R.W. 1983. Fibroblast immortality is a prerequisite for transformation by EJ c-*ras* oncogene. *Nature* **304:** 648–651.

12. Whyte P., Buchkovich J.J., Horowitz J.M., Friend S.H., Raybuck M., Weinberg R.A.,

and Harlow E. 1988. Association between an oncogene and an anti-oncogene: The adenovirus E1A proteins bind to the retinoblastoma gene product. *Nature* **334:** 124–129.

13. Hirakawa T. and Ruley H.E. 1988. Rescue of cells from *ras* oncogene-induced growth arrest by a second, complementing, oncogene. *Proc. Natl. Acad. Sci.* **85:** 1519–1523.

14. Serrano M., Lin A.W., McCurrach M.E., Beach D., and Lowe S.W. 1997. Oncogenic *ras* provokes premature cell senescence associated with accumulation of p53 and p16INK4a. *Cell* **88:** 593–602.

15. Lowe S.W. and Ruley H.E. 1993. Stabilization of the p53 tumor suppressor is induced by adenovirus 5 E1A and accompanies apoptosis. *Genes Dev.* **7:** 535–545.

16. Lowe S.W., Jacks T., Housman D.E., and Ruley H.E. 1994. Abrogation of oncogene-associated apoptosis allows transformation of p53-deficient cells. *Proc. Natl. Acad. Sci.* **91:** 2026–2030.

17. Lowe S.W., Ruley H.E., Jacks T., and Housman D.E. 1993. p53-dependent apoptosis modulates the cytotoxicity of anticancer agents. *Cell* **74:** 957–967.

Adenovirus early region 1A enables viral and cellular transforming genes to transform primary cells in culture

H. Earl Ruley

Cold Spring Harbor Laboratory, Cold Spring Harbor, PO Box 100, New York 11724, USA

The polyoma virus middle-T and the T24 Harvey ras1 *genes are individually unable to transform primary baby rat kidney cells. Adenovirus early region 1A provides functions required by these genes to transform primary cells following DNA-mediated gene transfer. These results suggest that separate establishment and transforming functions are required for oncogenic transformation of primary cells in culture.*

THE oncogenic DNA viruses encode proteins that are able to transform cells grown in tissue culture[1]. Virus-transformed cells, which can be isolated by their ability to form dense foci on monolayers of untransformed cells, typically retain and express all or some of the viral genes that are expressed early during lytic infection. Both primary and established cells can be transformed by polyoma virus (Py) and human adenovirus (Ad). However, the viral functions required to transform these cell types are different, at least in some cases.

Transformation of primary cells requires at least two separate functions. The first, an establishment function, is concerned with immortalization of cells, while the second, the transformation function, is required for full expression of an oncogenic phenotype[2-5]. Thus, establishment functions expressed by adenovirus early region 1A (E1A) or portions of the polyoma large-T antigen lead to the ability of primary cells to grow indefinitely in culture[2,3]. Additional functions expressed by adenovirus early region 1B (E1B) or the polyoma virus middle-T antigen result in phenotypic changes characteristic of oncogenic transformation[2-5] such as anchorage-independent cell growth and the ability to form tumours when cells are transplanted into syngeneic animals.

By contrast, transformation of established cell lines can require fewer viral functions. Thus, expression of the polyoma middle-T antigen alone is sufficient to transform a variety of cell lines[5]. Apparently, such cell lines constitutively express establishment functions that can substitute for those of the virus. The interaction between establishment and transforming functions is poorly understood, as is the mechanism by which they combine to elicit the transformed phenotype.

Recently, several genes have been isolated from cell lines established from human tumours, that have the ability to cause morphological transformation of the mouse NIH 3T3 cell line[6-10]. However, the ability of cellular oncogenes to transform cultured primary cells has not been critically addressed. Given the requirement for at least two viral functions for transformation of primary cells by polyoma and adenoviruses, it seemed quite likely that cellular oncogenes might also require additional functions. A failure to transform primary cells could be explained if oncogenes that score in the NIH 3T3 assay carry transforming functions but lack establishment functions.

0028-0836/83/330602—05$01.00

Kataoka T., Powers S., Cameron S., Fasano O., Goldfarb M., Broach J., and Wigler M. 1985. **Functional homology of mammalian and yeast RAS genes.** *Cell* **40:** 19–26. (Reprinted, with permission, from Elsevier ©1985.)

THE STUDY OF THE MOLECULAR BIOLOGY of human cancer—i.e., real human cancer, not cancer in model systems—began with the discovery of the first human oncogenes, many of which were identified by Mike Wigler's[1] group at Cold Spring Harbor in the early 1980s. These first human oncogenes turned out to be members of the *RAS* and *MYC* gene families. Although evolutionarily conserved homologs of *RAS* and *MYC* had been studied in rodent and avian systems, ideas about their cellular functions were in a very primitive state. Given the importance of human cancer, understanding their function in sufficient molecular detail became one of the most critical questions in the fields of oncology and molecular biology. The mere findings that Ras proteins were located in the inner surface of the plasma membrane and that Myc proteins were in the nucleus were of such significance at the time that they were reported in full-length articles in *Nature*. However, there were no good experimental approaches to really get at the cellular function of *RAS* or *MYC*, so new tactics had to be developed.

Two such novel approaches to explore *RAS* function were developed at Cold Spring Harbor in the mid 1980s. Jim Feramisco's[2] biochemical approach was to inject purified Ras protein directly into cells so that its immediate effects could be studied. Mike Wigler's approach was a pure genetics play. The idea was to study *RAS* genes in a eukaryotic system with powerful genetic tools so that genetics could be used to get at function. The fact that there was a vibrant yeast group at Cold

SCOTT POWERS holds a Ph.D. in biological sciences from Columbia University and, as a postdoctoral fellow in Mike Wigler's laboratory at Cold Spring Harbor in the early 1980s, used yeast to advance understanding of *ras* oncogene function. He was appointed as Director of the Genomics Division of Tularik Inc., a biotechnology company formed in 1995 to take advantage of a cancer gene discovery technology developed at Cold Spring Harbor. He rejoined Cold Spring Harbor Laboratory in 2004 as Director of the Cancer Genome Research Center and as an Associate Professor. powers@cshl.edu

Mike Wigler and Scott Powers, circa 1985. *(Photograph by Susan Lauter, courtesy of Cold Spring Harbor Laboratory Archives.)*

Spring Harbor at the time[3] played a big role in making the specific choice of yeast (versus *Drosophila*, for example).

Mike Wigler presented this idea to me on the first day of my postdoc in 1983. He also presented alternate projects, one studying alternative splicing of *KRAS*, which seemed remarkably uninspired in comparison. In thinking over the prospects of the different projects later on that night, I remembered Jim Watson's subtitle to a section in his second edition of *The Molecular Biology of the Gene* which went something like "Now is the time to seriously study yeast." I picked the yeast project.

The search for *RAS* genes in yeast was fun. I got to shuttle back and forth between Mike Wigler's Demerec lab with its intense East Coast atmosphere and the Delbrück yeast lab with its more laid back but wilder Western USA atmosphere. Jeff Strathern[4] would teach me yeast genetics at night with his golden retriever by his side and his green pick-up truck parked outside. He made yeast genetics seem both powerful and folksy at the same time. However, when Jeff switched into high gear and started talking about gene conversion and Holliday structures, yeast genetics just seemed way too complicated. Luckily, our collaborator and yeast geneticist Jim Broach[5] was gifted at being able to explain even the really complicated genetics in terms I could understand.

When we found out that Ed Scolnick's lab at Merck was also looking for *RAS* genes in yeast—and was ahead of us—our anxiety levels increased dramatically. Our scholarly three-pronged approach to find *RAS* genes in yeast had to end and

I focused all my efforts on the one that seemed most solidly grounded—to use human *RAS* genes to fish out yeast *RAS* genes by DNA hybridization. This got off to a bad start. I wasted several precious weeks screening other people's poorly constructed libraries. By complaining to Judy Abraham, another postdoc in Demerec, I elicited news of a magical library that had been constructed by the legendary yeast biologist Kim Nasmyth.[6] Although it took awhile to get an aliquot of his library, the first time I screened it, we were in like Flynn. This event marked the beginning of a scientific whirlwind. Slightly a year after I cloned yeast *RAS* genes, we had used genetics to determine the major normal function of *RAS* in yeast.

DNA sequencing showed that there were two yeast *RAS* genes and that the section of the corresponding proteins that were likely to interact with downstream effectors was remarkably conserved, in fact 100% conserved at the amino acid level! Mike and I were really excited by this and convinced ourselves that if we could find the downstream effector of *RAS* in yeast, it would be the same as or biochemically similar to an effector in human cells. The paper that is reproduced here—the third in a series of six *Cell* papers—addressed whether the human *RAS* genes could functionally substitute for the yeast *RAS* genes. It turns out they could, and this was the very first demonstration of functional homology between related genes from evolutionarily distant organisms. Jim Broach—who saw the historical importance of this more than any of us—pushed us along and did the lion's share of the genetic analysis. The work described in the paper really did bolster enthusiasm for our belief that if we could solve the mystery of what Ras proteins did in yeast, we would know what they did in human cells.

Were we ever wrong! As it turns out, the effector region of Ras proteins is 100% conserved in eukaryotic evolution, but the interacting effectors are not and they can have decidedly different biochemical functions. Why this is so has never been satisfactorily explained. Presumably, the reason why Ras proteins are so well conserved is that they interact with so many proteins that it becomes impossible to make amino acid changes without upsetting a whole collection of important interactions. And it is true that Ras proteins interact with at least two different effectors in yeast (neither of which are conserved in humans) and with a whole bunch of effectors in mammalian cells. The evolutionary theory for conservation of the Ras effector region would seem to demand that back in our evolutionary past, some primordial eukaryotic cell had a single Ras effector that has since vanished!

Despite our disappointment at not being able to solve the Ras problem in its entirety, we did make fundamental contributions to the subsequent elucidation of Ras protein function in human cells. We were the ones that discovered the first biochemical activator of Ras proteins (Cdc25) and determined the outline of how it worked as a GTP-GDP exchange factor to activate Ras proteins. This part of the Ras story—Cdc25-like proteins controlling Ras—is evolutionarily conserved between yeast and humans. In collaboration with Ira Herskowitz's laboratory, we also identified the genes responsible for the most important posttranslational modification of Ras proteins, farnesylation, which is also conserved between yeast and humans.

In a more general sense, the Kataoka et al. paper was really about pioneering a new type of molecular biology in which there was much more flexibility in picking the organism to study a given gene function. The fact that human *RAS* could substitute for yeast *RAS* surprised a lot of people and opened doors for scientists who didn't want to be put into a straightjacket of loyalty to a particular experimental species. A large number of people followed in our footsteps to study cancer gene function in genetically tractable organisms.

The boldness of being the first to "cross species" reflected Mike's inherent adventurousness and also reflected Jim Watson's encouragement for researchers at Cold Spring Harbor to take big chances. It is also worth noting that Jim Watson had created an open environment at the Laboratory that made it very easy—and even fun—to establish interactions between different laboratories.

Notes and References

1. Mike Wigler was appointed to head the Mammalian Cell Genetics Group at Cold Spring Harbor Laboratory in late 1978. He obtained his Ph.D. from the College of Physicians & Scientists of Columbia University, where he also carried out postdoctoral research.

2. See article by Dafna Bar-Sagi and Jim Feramisco in this volume (p. 69).

3. See article by Jim Hicks in this volume (p. 13).

4. Jim Strathern came to Cold Spring Harbor Laboratory in 1977 from Ira Herskowitz's lab to organize, with Jim Hicks, the nascent yeast group in Delbrück Laboratory. He left Cold Spring Harbor in 1984 to head a group at the Frederick Cancer Research Facility.

5. Jim Broach was a postdoctoral fellow and staff scientist at the Cold Spring Harbor Laboratory in the late 1970s. After moving to a faculty position at the nearby State University of New York at Stony Brook, he maintained a close working relationship with the Cold Spring Harbor yeast group. In 1984, he moved to the Molecular Biology Department of Princeton University, where he is now a Professor and Associate Director of the Lewis-Sigler Institute for Integrative Genomics.

6. Kim Nasmyth, who worked on fission yeast, came to Cold Spring Harbor in 1980 as a Robertson Fellow after a postdoctoral stint in Ben Hall's laboratory in Seattle. By 1983, when Scott Powers arrived in Cold Spring Harbor, Kim had already moved to the MRC Laboratory of Molecular Biology in Cambridge. In 1987, he took

a position at the Research Institute of Pathology in Vienna and more recently moved to Oxford University. The library constructed by Kim Nasmyth was described in: Nasmyth K.A. and Tatchell K. 1980. The structure of transposable yeast mating type loci. *Cell* **19:** 753–764.

Cell, Vol. 40, 19–26, January 1985, Copyright © 1985 by MIT

Functional Homology of Mammalian and Yeast *RAS* Genes

Tohru Kataoka,* Scott Powers,* Scott Cameron,*
Ottavio Fasano,* Mitchell Goldfarb,*†
James Broach,‡ and Michael Wigler*
* Cold Spring Harbor Laboratory
Cold Spring Harbor, New York 11724
‡ Department of Molecular Biology
Princeton University
Princeton, New Jersey 08544

Summary

Yeast spores lacking endogenous *RAS* genes will not germinate. If such spores contain chimeric mammalian/yeast *RAS* genes or even the mammalian H-*ras* gene under the control of the galactose-inducible *GAL10* promoter, they will germinate in the presence of galactose and produce viable haploid progeny dependent on galactose for continued growth and viability. These results indicate that the biochemical function of *RAS* proteins is essential for vegetative haploid yeast and that this function has been conserved in evolution since the progenitors of yeast and mammals diverged.

Introduction

The *ras* genes were first discovered as the oncogenes contained in the Harvey and Kirsten rat sarcoma viruses (Ellis et al., 1981). The H-, K-, and N-*ras* genes comprise a family of conserved mammalian genes that encode proteins of 188–189 amino acids (Taparowsky et al., 1983). A number of laboratories, using DNA-mediated gene transfer, have demonstrated the presence of "activated" *ras* genes in the DNA of many tumor cells. These *ras* genes contain missense mutations which render them capable of morphological and tumorigenic transformation of NIH3T3 cells, an established cell line of murine origin (Tabin et al., 1982; Reddy et al., 1982; Taparowsky et al., 1982, 1983; Shimizu et al., 1983; Yuasa et al., 1983; Capon et al., 1983). The mammalian *ras* proteins are synthesized initially as cytoplasmic precursors that undergo a processing event while becoming localized to the inner surface of the plasma membrane (Shih et al., 1982b). Both mutant and normal *ras* proteins bind guanine nucleotides with high affinity (Shih et al., 1980, 1982a; Papageorge et al., 1982). The normal *ras* proteins, but not the mutant proteins, also have weak GTPase activity (Sweet et al., 1984; McGrath et al., 1984). Little else is known about the biochemical function of the mammalian *ras* proteins.

The yeast Saccharomyces cerevisiae contains two genes, *RAS1* and *RAS2*, which encode proteins that are highly homologous to the mammalian *ras* proteins, particularly in their N-terminal domains (Defeo-Jones et al., 1983; Powers

et al., 1984; Dhar et al., 1984). Although neither *RAS1* nor *RAS2* are by themselves essential genes, at least one functional *RAS* gene is required for the resumption of vegetative growth by haploid yeast spores (Kataoka et al., 1984; Tatchell et al., 1984). This observation has allowed us to test for functional homology between the mammalian and yeast *RAS* proteins. To this end, we have made several mammalian/yeast *RAS* gene constructions in vitro and tested them for their ability to complement the loss of endogenous *RAS* function in yeast.

Results

An N-Terminal Chimeric Gene Complements *ras1⁻ ras2⁻* Yeast

The proteins encoded by the yeast *RAS2* and the human H-*ras* genes are most similar in their N-terminal domains (Defeo-Jones et al., 1983; Powers et al., 1984; Dhar et al., 1984). The yeast *RAS2* protein contains seven initial amino acids not present in the H-*ras* protein. After that, there is close to 90% homology between the proteins for the next 80 amino acid positions. There is nearly 50% homology for the 80 amino acid positions which follow that, and then homology breaks down in what we have called the variable domain. Finally, there is homology again in the four C-terminal amino acid positions. The yeast *RAS2* protein is predicted to be 322 amino acids, much larger than the 189 amino acid H-*ras* protein, mainly because of a larger variable region.

There is a conserved cleavage site for the restriction endonuclease Mst I in both the H-*ras* and *RAS2* genes, corresponding to the 73rd and 80th amino acid positions of the respective proteins. We have utilized this site to create a chimeric gene that encodes the first 73 amino acids of H-*ras* and the remaining 242 amino acids of *RAS2*. The resulting protein differs from the first 79 positions of the H-*ras* at only one position, encoding asparagine instead of threonine at position 74. It differs from *RAS2* at 17 positions in this region. This chimeric protein resembles the normal mammalian protein at all positions at which amino acid substitutions are known to activate the transforming potential of the mammalian protein. These coding regions have been arranged as part of a transcription unit that utilizes the *GAL10* galactose-inducible promoter (St. John and Davis, 1981; Broach et al., 1983) and *RAS2* termination sequences. The entire transcription unit is contained on the plasmid pTKJB-2(gly), which, in addition, contains the origin of replication and the β-lactamase gene of pBR322, and the *LEU2* gene of yeast. The details of this construction are given in Figure 1. A similar plasmid, pTKJB-2(val), was constructed differing only in encoding valine at position 12 of the hybrid H-*ras*/*RAS2* protein. We also constructed plasmid pTKJB-3(gly) which places the entire yeast *RAS2* coding sequence under the control of the GAL10 promoter (see Figure 1).

DNAs from pTKJB-2(gly) and pTKJB-2(val) were cleaved with Eco RI, which cuts within the *LEU2* gene, and used

---

†Present address: Department of Biochemistry, Columbia University College of Physicians and Surgeons, New York, NY 10027.

Hanahan D. 1985. **Heritable formation of pancreatic β-cell tumours in transgenic mice expressing recombinant insulin/simian virus 40 oncogenes.** *Nature* **315**: 115–122. (Reprinted, with permission, from Macmillan Publishers Ltd ©1985.)

THE PAPER DESCRIBES LINEAGES OF TRANSGENIC MICE that reproducibly developed cancer by virtue of expressing an oncogene selectively in a rare cell type: the insulin-producing β cell of the pancreatic islets. The hybrid oncogene was composed of the 5′ regulatory region of a rat insulin gene, aligned to transcribe the oncogene-encoding early region of the DNA tumor virus SV40. At the time, the work was significant in three respects:

- It established proof of principle that gene expression could be targeted to a rare and important cell type, the β cell that produces insulin and mediates glucose homeostasis, using a relatively small segment of the insulin gene.

- It demonstrated that an oncogene, originally identified by its ability to transform cells in culture, was able when carried in the mouse germ line to elicit the development of tumors arising out of normal cells resident in their natural tissue microenvironment. These and other transgenic onco-mice discussed below presented a strong proof of principle that oncogenes could originate, rather than just promote, malignancies.

- It confirmed that carcinogenesis was a multistep event. The anatomy of the cell type being targeted, localized into 400 natural focal nodules (the islets of Langerhans), presented a particularly convincing case for the multistep nature of carcinogenesis. Only a few of the islets expressing the SV40 oncogenes pro-

---

DOUG HANAHAN is currently an American Cancer Society Research Professor in the Department of Biochemistry and Biophysics at the University of California at San Francisco. His Ph.D. research during 1978–1983 was carried out jointly at Cold Spring Harbor Laboratory and Harvard University, where his supervisor was Paul Doty. He was a Harvard Junior Fellow from 1981 to 84. From 1983 onward, he held progressively more senior staff appointments at Cold Spring Harbor Laboratory before moving to a tenured position at the University of California at San Fransisco in 1988. dh@biochem.ucsf.edu

gressed into tumors, suggesting that the oncogene was necessary but insufficient to manifest a tumor. Importantly, the oncogene was demonstrably expressed well before tumors formed, and most of the islets in older mice showed signs of aberrant proliferation, suggesting that (1) the oncogene was inducing β-cell hyperproliferation and consequent development of premalignant lesions and (2) additional alterations (events) potentially independent of the oncogene were necessary to produce a full-blown tumor.

The work that generated these conclusions had begun in 1983, 2 years before the *Nature* paper was published, when I learned the techniques for generating transgenic mice. In January 1984, I began the microinjections that would produce the founder animals of three RIP-Tag lines, as well as a variant line (RIR-Tag). The founder mice all died with pancreatic tumors between April and June 1984. By then, two other groups had already generated transgenic mice expressing oncogenes: Richard Palmiter and Ralph Brinster, leaders in developing transgenic mouse technology, had made mice expressing the same Tag oncogene in the brain epithelium, eliciting choroid plexus tumors.[1] Then, Tim Stewart and Phil Leder reported that transgenic mice expressing the cellular c-*myc* oncogene in the mammary glands under control of the mouse mammary tumor virus regulatory region heritably developed breast cancer.[2] Both models similarly confirmed the causality of oncogenes, and the Stewart/Leder model also implicated a multistep pathway of tumorigenesis, in that only one of the ten oncogene-expressing mammary glands developed tumors, suggestive of other necessary events. Additional models were reported later: the Eμ-Myc model of B-cell lymphoma[3] and a model of bone cancer induced by the *fos* oncogene.[4] Collectively these papers and these models established a foundation for a field that has continued to flourish and develop for 20 years,[5] becoming a major force in modern cancer research both for investigating mechanisms of tumorigenesis and, increasingly, for testing mechanism-guided therapeutic strategies.

My entrée into transgenic mouse technology in 1983 could have been perceived as presumptuous and certainly risky and questionable for a lone scientist working at an institution with no relevant expertise. The technique was then difficult, and the prospects for achieving correct tissue-specific expression were uncertain. Palmiter and Brinster were predominant and viewed as having extraordinary expertise, with a stream of high-profile publications reporting expression and phenotypes (ranging from cancer to "big mice"), but using promoters that were ubiquitously or very widely expressed in many different types of cells. The one prior attempt with a cell-type-specific promoter, derived from a hemoglobin gene by Frank Costantini and Liz Lacy, produced transgenic mice that failed to express properly,[6] raising concerns (subsequently dispelled by many, including myself) that tissue-specific genes might not be so amenable to the transgenic gene transfer technology. But with the support and encouragement of Jim Watson and Joe Sambrook at Cold Spring Harbor, I fearlessly dove in, brashly confident that I would manage to succeed.

A review of the work in 1985 raised the issue that viral oncogenes such as those encoded by SV40 and adenovirus were poor choices because neither virus caused cancer in its natural host. The question was whether the RIP-Tag model was relevant to naturally occurring cancers, given that SV40 T antigen was not a cellular oncogene like Ras or Myc. Although I could not at the time convincingly rebut such philosophical critiques, it was my intuition that the ontogeny of tumorigenesis elicited by T antigen was likely to be instructive about the principles of cancer. Later, Ed Harlow and colleagues discovered that an adenoviral oncogene (and subsequently SV40 T anti-

Doug Hanahan, 1988. *(Photograph by Susan Lauter, courtesy of Cold Spring Harbor Laboratory Archives.)*

gen) inactivated the retinoblastoma tumor suppressor pathway,[7] and others showed that p53 was similarly inactivated by Tag and other viral oncogenes.[8] In this light, T antigen could be viewed as a means to knock out in mice what have proved to be the predominant tumor suppressor pathways abrogated in one way or another in most forms of human cancer.

I did not imagine in 1985 that my lab and others would still be studying the RIP-Tag mice in 2007. Remarkably, however, the RIP-Tag models have during the past two decades taught us valuable lessons about the functional importance and regulatory mechanisms of hallmark parameters of cancer, including angiogenesis and the tumor vasculature; survival signaling and resistance to apoptosis, invasion and metastasis; self-tolerance and tumor immunity; telomere maintenance; and extracellular proteases and their supply by infiltrating leukocytes. In many ways, the value of the models has continued to increase as new uses for them have been found. For example, we have developed stage-specific therapeutic trial designs using the RIP-Tag2 model that are suggesting new strategies for multitargeting human cancers, aiming with mechanism-based drugs at different stages of progression and at the distinct cell types that together constitute the tumor organ.

Notes and References

1. Brinster R.L., Chen H.Y., Messing A., van Dyke T., Levine A.J., and Palmiter R.D. 1984. Transgenic mice harboring SV40 T-antigen genes develop characteristic brain tumors. *Cell* **37:** 367–379.

2. Stewart T.A., Pattengale P.K., and Leder P. 1984. Spontaneous mammary adenocarcinomas in transgenic mice that carry and express MTV/myc fusion genes. *Cell* **38:** 627–637.

3. Adams J.M., Harris A.W., Pinkert C.A., Corcoran L.M., Alexander W.S., Cory S., Palmiter R.D., and Brinster R.L. 1985. The c-*myc* oncogene driven by immunoglobulin enhancers induces lymphoid malignancy in transgenic mice. *Nature* **318:** 533–538.

4. Ruther U., Komitowski D., Schubert F.R., and Wagner E.F. 1989. c-*fos* expression induces bone tumors in transgenic mice. *Oncogene* **4:** 861–865.

5. Hanahan D., Wagner E.F., and Palmiter R.D. 2007. The origins of oncomice: A history of the first transgenic mice genetically engineered to develop cancer. *Genes Dev.* **15:** 2258–2270.

6. Lacy E., Roberts S., Evans E.P., Burtenshaw M.D., and Costantini F.D. 1983. A foreign β-globin gene in transgenic mice: Integration at abnormal chromosomal positions and expression in inappropriate tissues. *Cell* **34:** 343–358.

7. Whyte P., Williamson N.M., and Harlow E. 1989. Cellular targets for transformation by the adenovirus E1A proteins. *Cell* **56:** 67–75.

8. Ahuja D., Saenz-Robles M.T., and Pipas J.M. 2005. SV40 large T antigen targets multiple cellular pathways to elicit cellular transformation. *Oncogene* **24:** 7729–7745.

Reprinted from Nature, Vol. 315, No. 6015, pp. 115-122, 9 May 1985
© *Macmillan Journals Ltd., 1985*

Heritable formation of pancreatic β-cell tumours in transgenic mice expressing recombinant insulin/simian virus 40 oncogenes

Douglas Hanahan

Cold Spring Harbor Laboratory, Cold Spring Harbor, New York 11724, USA

Following the transfer into fertilized mouse eggs of recombinant genes composed of the upstream region of the rat insulin II gene linked to sequences coding for the large-T antigen of simian virus 40, large-T antigen is detected exclusively in the β-cells of the endocrine pancreas of transgenic mice. The α- and δ-cells normally found in the islets of Langerhans are rare and disordered. Well-vascularized β-cell tumours arise in mice harbouring and inheriting these hybrid oncogenes.

INSULIN, a polypeptide hormone involved in the control of carbohydrate metabolism, is synthesized and stored in the β-cells of the endocrine pancreas, from which it is released into the bloodstream in response to various signals. The endocrine pancreas is comprised of isolated islands or islets, which are dispersed throughout the larger mass of the exocrine pancreas; the islets consist of four major cell types, α, β, δ and PP, which synthesize the hormones glucagon, insulin, somatostatin and pancreatic polypeptide, respectively[1,2]. Rat and human insulin genes have been cloned and analysed[3-13]. There are two non-allelic rat insulin genes that differ in nucleotide sequence and in the number of introns; the rat insulin I and II genes have one and two introns, respectively[5,6]. The two genes share a region of considerable sequence homology in the 5' flanking region, extending ~300 base pairs (bp) upstream of the point of transcriptional initiation[12]; the human gene is 75% homologous over the first 240 bp of this region[8]. Expression of hybrid genes using the 5' flanking regions of both the human insulin and rat insulin I genes occurs when they are introduced transiently into cultured cells derived from an insulinoma[11].

Here I seek to address aspects of insulin gene expression by the introduction of recombinant insulin genes into the mouse germ line, using microinjection of DNA into fertilized mouse embryos. This technique has been used previously to examine the control and consequences of expression of a number of genes[14-23], including two oncogenes. The complete simian virus 40 (SV40) early region, including the viral transcriptional enhancer and promoter, is not expressed at detectable levels in normal tissues of transgenic mice harbouring it, but choroid plexus tumours heritably arise, and these tumours produce high levels of large-T antigen[19]. The cellular proto-oncogene c-*myc*, when linked to the long terminal repeat of mouse mammary tumour virus, is expressed in a number of tissues, including mammary cells, and two lines of mice heritably develop breast tumours[20].

The recombinant oncogenes used here consist of regulatory information associated with the rat insulin II gene linked to protein coding information for the oncogene SV40 large-T antigen (Tag). The rationale for using such a hybrid is threefold: (1) to examine the ability of the insulin sequences to mediate correct tissue- and cell-type specific expression in the β-cells of the endocrine pancreas, using a readily identified and distinguishable marker protein (a viral antigen); (2) to examine the consequences to the organism of tissue-specific expression of

Fig. 1 Structure of recombinant insulin/large-T antigen genes. The plasmids pRIP1–Tag (*a*) and pRIR–Tag (*b*) consist of the coding information of the SV40 early region fused to sequences derived from the 5' flanking region of the rat insulin II gene. Boxed regions denote insulin gene flanking DNA and the thick solid line refers to the SV40 early region. The cap site where insulin gene transcription initiates is indicated by an L-shaped arrow, and the associated transcriptional enhancer element[11] (EH) is shown (the insulin gene promoter, which lies between them, is not indicated). The points of SV40 large-T antigen translation initiation and transcription termination are indicated by boxed ATG and A$_n$ respectively. The *Bgl*I site in SV40 was converted to an *Xba*I site using an oligonucleotide linker, and the *Xba*/*Bgl*I to *Bam* fragment comprising the SV40 early region[24] was linked to two different orientations of the insulin gene flanking region. This fragment of the SV40 early region includes the early mRNA cap sites, but lacks all known components of the SV40 early promoter. For *RIP1*-Tag, an *Xba* linker was inserted into the *Dde*I site at +8 bp relative to the point of initiation of insulin gene transcription, and the *Bam* to *Dde*/*Xba* fragment from −660 to +8 was combined with the *Xba*/*Bam* fragment of the SV40 early region and inserted as a *Bam* linear into a derivative of pBR322 that lacks the R1 site, and includes a *lac*UV5 promoter in place of the R1 to *Bam* fragment of pBR. For the plasmid pRIR–Tag, which carries the insulin flanking region in the opposite orientation, a *Dde* site at −520 was converted to an *Xba* site, and the *Bam*/*Xba* fragment extending from +180 nucleotides past the cap to −520 bp upstream was combined with the structural gene for large-T antigen and inserted into the same plasmid vector. The two hybrid genes were similarly prepared for microinjection, except that for *RIR*-Tag, the *Bam* insert was first purified from the plasmid vector by gel electrophoresis and electroelution, followed by passage over a DEAE–Sephacel column and pRIP1–Tag was linearized using *Sal*I. The DNAs were extracted with phenol, chloroform, precipitated in 70% ethanol, resuspended in 10 mM NaCl, 5 mM Tris pH 7.4, 0.1 mM EDTA and passed through a 1-ml spin column of Sepharose CL6B, after which the DNA was diluted to concentrations of ~75 copies per pl for pRIP1–Tag/*Sal* or ~20 copies per pl for the *RIR–Tag*/*Bam* insert. The DNAs were injected into fertilized one-cell embryos essentially as described[14,16], except that the injected embryos were cultured overnight to the two-cell stage before being re-implanted into the oviducts of pseudopregnant females. The F$_2$ embryos were derived from matings of B6D2F1 (C57BL/6J×DBA/2J) males and females, obtained from the Jackson laboratory.

a

RIP1–Tag

b

RIR–Tag

Franza B.R. Jr., Rauscher F.J. 3rd, Josephs S.F., and Curran T. 1988. **The Fos complex and Fos-related antigens recognize sequence elements that contain AP-1 binding sites.** *Science* **239**: 1150–1153. (Reprinted, with permission, from AAAS ©1988.)

I N A SMALL ROOM—OR AS SOME CHARACTERIZED IT, a small walk-in closet—I spent many wonderful days studying biomolecules and their interactions. The room was located on the second floor of the northwest corner of a building that shares many features of the famed Stazione Zoologica in Naples and that in 1971 was renamed in honor of the revered scientist, Barbara McClintock. My usual routine was to arrive by 5:30 a.m., typically when Winship Herr would be leaving James Laboratory after another all-nighter, make coffee, glance at the view of the Harbor, and then get down to work. In the winter and spring of 1987, in this very same room, I invented and developed the microscale DNA affinity precipitation (DNAP) assay. I had decided to use the resolution power of two-dimensional gel chromatography to observe and quantify proteins that bound, with varying specificity to distinct, short DNA structures. The results have proven informative and challenging.

Among the first DNA sequence elements that we investigated were the targets for two transcription factors: NF-κB and AP-1. We mixed short synthetic oligonucleotides containing the binding sites for NF-κB and AP-1 with extracts of cellular proteins and, after recovery, resolved the molecules that associated specifically with the oligonucleotide probes. The results of the first few assays were truly exhilarating. What we observed were several proteins, some of which we were already quite familiar, associated with the AP-1 target oligonucleotide. All of these proteins were distinct from those associating with the NF-κB target. Specificity was determined by comparison of protein binding to displays generated by oligonucleotide structures differing in sequence by as little as 1 base pair.

BOB FRANZA worked at Cold Spring Harbor during the 1980s and early 1990s, first as a postdoctoral fellow and later as a member of the scientific staff. He is currently Co-founder and Executive Scientific Director of the Seattle Science Foundation and Affiliate Professor of Bioengineering, University of Washington, Seattle. BobF@seattlesciencefoundation.org

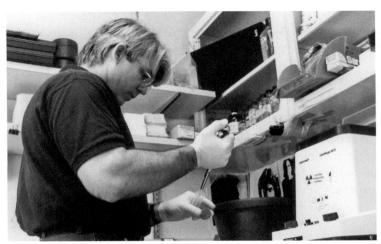

Bob Franza, circa 1990s. *(Photograph by Margot Bennett,
courtesy of Cold Spring Harbor Laboratory Archives.)*

Today, instead of photocopying results from the pages of my three-ring note-
books and mailing them to my collaborators, I could have much more efficiently
communicated via the Internet with my co-authors, Steven Josephs, Tom Curran,
and Frank Rauscher, the excitement that we all shared in October and November
of 1987—excitement that lasted long after our paper was published in *Science* in
March of 1988. But I'll always remember standing in that little room, looking out
over Cold Spring Harbor and talking on the phone with Tom, Frank, and Steve
about our discoveries. We thought them important at the time and they have also
formed the basis of both ongoing collaborations and continued friendships.

I live in the Pacific Northwest now; but I often wish I could have simply continued, in that special room, to probe the dynamic network of molecular interactions that underlie the adaptive control systems called "cells."

In the intervening years I've continued to be challenged by the meaning of our
results and this has motivated me to invent and to build collaborations with other
inventors in two areas:

- One is the creation of a language that enables unambiguous, immediate, and ver-
  ifiable description of spatially contextualized, dynamic interactions of different
  biomolecules.

- The other is the development of a plan for an integrated, multimodal imaging
  platform, in honor of Lyman Spitzer Jr.—the driving force behind the Hubble
  telescope. We affectionately call it the "Biologists' Hubble." The platform, combined with computational simulation, will enable hypotheses of the dynamics

and spatial distribution of molecular interactions in living systems to be modeled, simulated, and probed. Access via digital networks will ensure the broadest participation of individuals as well as teams of scientists.

During the 12 years since I departed Cold Spring Harbor Laboratory, I've had the privilege to collaborate with a small but tremendously inventive group of colleagues. To that end, we established in 2006 the not-for-profit Seattle Science Foundation whose one sentence "elevator pitch" is: "SSF creates and nurtures social networks of experts to solve major challenges in science and medicine." By reducing the barriers to creativity and facilitating effective collaboration, we hope to develop mechanisms by which individuals interested in understanding living systems can express their ideas, simulate the properties of their models, share quantitative observations efficiently, and educate their peers and other students, thereby contributing to the well-being of humanity.

The Fos Complex and Fos-Related Antigens Recognize Sequence Elements That Contain AP-1 Binding Sites

B. Robert Franza, Jr.,* Frank J. Rauscher III,
Steven F. Josephs, Tom Curran

The Fos protein complex and several Fos-related antigens bind directly or indirectly to a common sequence element that is similar to the consensus binding site for HeLa cell activator protein 1 (AP-1). This element is present in a negative regulatory sequence in the differentiation-sensitive adipocyte gene, aP2; in a transcriptional enhancer for the Gibbon ape leukemia virus; and in a region of the human immunodeficiency virus (HIV) long terminal repeat partially characterized as a negative regulatory element. The protein level and binding activity of Fos and Fos-related antigens increase rapidly after calcium ionophore treatment of a CD4+ human lymphoblast cell line, H9. These data suggest that several proteins may associate with the AP-1 binding site. Moreover, temporally regulated control of the level of each protein could represent a mechanism for modulation of these putative mediators of gene expression.

THE PROTO-ONCOGENE c-*fos* ENcodes a nuclear phosphoprotein (Fos) that is associated with chromatin and that displays DNA binding in vitro (*1*). Fos is expressed at relatively low levels in the majority of cell types; however, it can be rapidly and transiently induced by a great variety of extracellular stimuli (*2*). It has been proposed that Fos might function as a nuclear "third messenger" molecule in coupling short-term signals elicited by cell-surface stimulation to long-term alteration in cellular phenotypes by regulating expression of specific target genes (*3*). Indeed, v-*fos* has been shown to stimulate transcription of selected promoters in trans (*4*). Recently, evidence was presented that Fos participates in a nuclear protein complex with a sequence element in a control region of the adipocyte (3T3-F442A) differentiation-sensitive gene, aP2 (*5*). We noticed considerable similarity between the results of oligonucleotide gel retention assays done with the fat-specific element 2 (FSE2) by Distel *et al.* (*5*) to those of Quinn *et al.* (*6*) with an enhancer element from the Gibbon ape leukemia virus (GALV) long terminal repeat (LTR). These two oligonucleotides share a sequence motif within the region (−357 to

−278) of the HIV-LTR that acts as a negative regulatory element (NRE) (*7*). This common sequence element is remarkably similar to the consensus binding site AP-1 (*8–10*) (see Fig. 1). To investigate a possible association of Fos with these nucleotide sequences we have applied an assay system that permits direct analysis of cellular proteins that interact with specific nucleic

Fig. 1. Sequences of synthetic oligonucleotides used in DNAP assays. DFSE2 is a two-site iteration of the aP2 gene control element (bases −122 to −98 from transcription start site) as described by Distel *et al.* (*5*). DMFSE2 is a two-site iteration of a mutant of FSE2. The sites of mutation are −111 and −109. This mutant was designed to minimally disrupt the AP-1 site of FSE2. DGALVEN is a two-site iteration of the region identified in the GALV Seato strain LTR as sufficient for enhancer activity by Quinn *et al.* (*6*). HIV −(357/316) is a single iteration of these bases in the HIV-LTR as described previously (*7*). The * designates bases of significant sequence similarity to the reported AP-1 consensus (*8–10*). The underlined bases indicate the two-base insertion between the two sequence elements. Oligonucleotides were synthesized, purified, and biotinylated as described (*11*). Under the complementary strand arrows indicate the 5' to 3' orientation of the AP-1 related sequences.

B. R. Franza, Jr., Cold Spring Harbor Laboratory, Cold Spring Harbor, NY 11724.
F. J. Rauscher III and T. Curran, Roche Institute of Molecular Biology, Hoffmann–La Roche, Inc., Nutley, NJ 07110.
S. F. Josephs, Laboratory of Tumor Cell Biology, National Cancer Institute, National Institutes of Health, Bethesda, MD 20892.

*To whom correspondence should be addressed.

Whyte P., Buchkovich K.J., Horowitz J.M., Friend S.H., Raybuck M., Weinberg R.A., and Harlow E. 1988. **Association between an onco-gene and an anti-oncogene: the adenovirus E1A proteins bind to the retinoblastoma gene product.** *Nature* **334**: 124–129. (Reprinted, with permission, from Macmillan Publishers Ltd ©1988.)

A LTHOUGH THEY SATISFY THEIR NEEDS in different ways, all viruses rely heavily on their host cells for completion of their life cycles. The biochemical inter-face where a viral protein interacts with a host process has proven to be a useful place to look not only for important events of the virus' life cycle, but also, in some remarkable ways, for key features of host metabolism.

Our knowledge of how small DNA viruses such as polyomavirus, SV40, and adenoviruses transform mammalian cells is still incomplete, but the principles that underlie their actions are now well framed. These viruses are small enough that they are obliged to rely heavily on the host's enzymes to replicate their genomes. To commandeer the DNA replication machinery, these viruses have evolved clever strategies to drive reluctant host cells into the S phase of the cell cycle and to pro-tect the infected cells from fail-safe mechanisms that block synthesis or kill cells that enter S phase inappropriately. When viral oncoproteins induce entry into S phase in circumstances where the viral life cycle is unable to be completed, the com-mandeered cells are stimulated to divide at inappropriate times, becoming immor-tal in cell culture, thus clearing one of the major hurdles to cell transformation.

Among the key mechanisms that viruses use to drive infected cells into DNA synthesis is inactivation of the cell cycle regulators that block untimely entry into S phase. The interaction of the adenovirus-coded E1A proteins with pRB, the prod-

ED HARLOW is currently Professor and Chair, Department of Biological Chemistry and Molecular Pharmacology, Harvard Medical School. Ed did his undergraduate training at the University of Oklahoma and received his Ph.D. from the Imperial Cancer Research Fund Laboratories in 1982. Much of his work on the function of viral oncoproteins was performed while on the staff at Cold Spring Harbor Laboratory where in 1988 he and his laboratory discovered how small DNA tumor viruses alter cell cycle control by synthesizing viral proteins that interact with and inactivate neg-ative regulators of proliferation. eharlow@hms.harvard.edu

Ed Harlow, 1994. *(Photograph by Marléna Emmons, courtesy of Cold Spring Harbor Laboratory Archives.)*

uct of the cell-coded retinoblastoma tumor suppressor gene, is one of the most helpful in explaining how the virus manages to subvert the host cell cycle. pRB normally acts as a brake to control the progression of the cell from the G_1 phase of the cell cycle into the S phase. The viral E1A proteins bind to pRB, inactivate it, and allow cells to begin DNA synthesis. The concept of viral proteins binding to cellular negative regulators, often tumor suppressor proteins, and blocking their normal roles was first established in the paper by Whyte et al., which is reproduced on the enclosed DVD. The essential work that followed from many laboratories showed how pRB worked and how the adenovirus proteins were able to neutralize the pRB function. Similar mechanisms used by many other DNA viruses probably explain a good portion of the transforming roles of the DNA virus' oncoproteins. Analogous interactions between cellular positive and negative regulators are also now well-documented.

The paper by Whyte et al. stands directly on the shoulders of earlier work. The complexes formed between cellular p53 and SV40 large T antigen,[1,2] cellular Src and polyomavirus middle T antigen,[3] and p53 and adenovirus E1B protein[4] were well-studied examples of physical interactions of viral oncoproteins with cellular polypeptides. In some cases, it was already suspected that viral transformation relied on these interactions. What was missing was a clear understanding of why these complexes were formed between viral and host proteins. The normal roles of the cellular proteins needed to be clarified.

The intellectual and technical buildup to the work in this paper is straight-forward. A panel of monoclonal antibodies specific for the adenovirus E1A protein had been prepared, and during the characterization of these antibodies, it became clear that the E1A protein did not float freely in the cell but was bound to a series of cellular proteins of unknown functions. Determining what these proteins were and what roles they played became a major goal of the Harlow laboratory. The 105-kD protein was a prime target for study. The protein was present in high abundance in cells and became one of our main targets for purification and characterization. Since protein purification can be a lengthy and frustrating business, we also searched the scientific literature for previously described cellular proteins with a molecular mass of approximately 105,000 and potentially relevant functions. Our hope was to short-circuit the more laborious route of protein purification and sequencing. This fast-track strategy proved to be successful when, in late 1987, Wen Hwa Lee's laboratory published the characterization of the retinoblastoma gene product.[5] Their paper came out in *Nature* while I was in London working on the first edition of *Antibodies: A Laboratory Manual*. Reading this paper in the library of the Imperial Cancer Research Fund in South Mims, it was clear to me that pRB was a strong potential target for interaction with the adenovirus E1A protein. But this thought was not mine alone. By the time I returned to Cold Spring Harbor in the following week, Peter Whyte and Karen Buchkovich in our laboratory independently had the same idea and had already begun the key experiments. As a laboratory head, you have to be delighted with that kind of recognition, insight, and initiative from your graduate students. Twenty years later, I am still proud to tell that story.

The experiments that followed were easy for us since we were expert at using antibodies to detect protein complexes and then to identify the various components of the complexes. Our manuscript was reviewed quickly during the Cold Spring Harbor Symposium in early June 1988, and the paper was published in *Nature* a few weeks later. I learned later that as we were presenting the data at the Symposium, the manuscript was being reviewed in other places on campus. We owe a debt to Jim Watson, who without my knowledge, argued, effectively it turns out, with the editors of *Nature* that the paper should be published as an article rather than a letter as they were thinking.

During the hot Cold Spring Harbor summer, we collected data, published later that year,[6] showing that the precise regions of E1A needed to bind to pRB were also required for the protein's transforming ability. The story fit so beautifully: Binding to pRB was needed for viral transformation, and pRB had a role that explained E1A's action.

Another important anecdote associated with the publication of the Whyte *Nature* paper stems not from our group but a related manuscript from David Livingston's laboratory on the interaction of SV40 large T antigen with pRB. They made this discovery independently from our work but were prompted to begin the studies because of earlier presentations from our group. With the greatest of collegial manners, David requested that *Cell* not publish their work until our paper came out in *Nature*. *Nature* published our paper on July 14; *Cell* published theirs on July 15.[7] And in the introduction to their paper, they made a gracious reference to the priority of our work. This remains for me one of the finest examples of how competitive scientists can work hard, achieve success, share credit, and still keep the history books straight. Throughout the following years,

David's laboratory and mine have worked hard to maintain both the competitive productivity and the collegial standards set by his group. I will always be grateful for the example he set.

Notes and References

1. Lane D.P. and Crawford L.V. 1979. T antigen is bound to a host protein in SV40-transformed cells. *Nature* **278**: 261–263.

2. Linzer D.I. and Levine A.J. 1979. Characterization of a 54K dalton cellular SV40 tumor antigen present in SV40-transformed cells and uninfected embryonal carcinoma cells. *Cell* **17**: 43–52.

3. Courtneidge S.A. and Smith A.E. 1983. Polyoma virus transforming protein associates with the product of the c-src cellular gene. *Nature* **303**: 435–439.

4. Sarnow P., Ho Y.S., Williams J., and Levine A.J. 1982. Adenovirus E1b-58kd tumor antigen and SV40 large tumor antigen are physically associated with the same 54 kd cellular protein in transformed cells. *Cell* **28**: 387–394.

5. Lee W.H., Shew J.Y., Hong F.D., Sery T.W., Donoso L.A., Young L.J., Bookstein R., and Lee E.Y. 1987. The retinoblastoma susceptibility gene encodes a nuclear phosphoprotein associated with DNA binding activity. *Nature* **329**: 642–645.

6. Whyte P., Ruley H.E., and Harlow E. 1988. Two regions of the adenovirus early region 1A proteins are required for transformation. *J. Virol.* **62**: 257–265.

7. DeCaprio J.A., Ludlow J.W., Figge J., Shew J.Y., Huang C.M., Lee W.H., Marsilio E., Paucha E., and Livingston D.M. 1988. SV40 large tumor antigen forms a specific complex with the product of the retinoblastoma susceptibility gene. *Cell* **54**: 275–283.

Association between an oncogene and an anti-oncogene: the adenovirus E1A proteins bind to the retinoblastoma gene product

Peter Whyte[*†], Karen J. Buchkovich[*], Jonathan M. Horowitz[‡], Stephen H. Friend[‡§], Margaret Raybuck[*†], Robert A. Weinberg[‡] & Ed Harlow[*‖]

* Cold Spring Harbor Laboratory, Cold Spring Harbor, New York 11724, USA
‡ Whitehead Institute for Biomedical Research, Cambridge, Massachusetts 02142, USA and Department of Biology, Massachusetts Institute of Technology, Cambridge, Massachusetts 02139, USA
§ Division of Hematology-Oncology, The Children's Hospital, Dana-Farber Cancer Institute, Department of Pediatrics, Harvard Medical School, Boston, Massachusetts 02115, USA

One of the cellular targets implicated in the process of transformation by the adenovirus E1A proteins is a 105K cellular protein. Previously, this protein had been shown to form stable protein/protein complexes with the E1A polypeptides but its identity was unknown. Here, we demonstrate that it is the product of the retinoblastoma gene. The interaction between E1A and the retinoblastoma gene product is the first demonstration of a physical link between an oncogene and an anti-oncogene.

REGULATION of cellular proliferation is a complex process that involves both positively and negatively acting signals. Tumourigenesis results from alterations in genes whose protein products are involved in these signalling pathways. The DNA tumour viruses encode a set of proteins that are capable of overriding and reprogramming normal regulation of cellular

† Present addresses: Fred Hutchinson Cancer Center, 1124 Columbia Street, Seattle, Washington 98104, USA (P.W.) and Amersham International plc., Forest Farm Industrial Estate, Whitchurch, Cardiff, UK (M.R.).
‖ To whom correspondence should be addressed.

growth; consequently, they have been widely used as model systems for studying cellular transformation. The oncogenes—tumour-inducing genes—from polyomavirus, simian virus 40 (SV40) and adenovirus are able to induce a number of distinct changes in cell phenotype, including immortalization, secretion of growth factors, loss of contact inhibition, anchorage-independent growth and morphological transformation. Unlike the transforming retroviruses, these DNA viruses contain oncogenes that do not appear to have cellular homologues. Although functional similarities have been shown between cellular oncogenes

In part, the apocalyptic sense in the late 1960s that molecular biology's Golden Age had past reflected a commonly held but flawed view that all important scientific breakthroughs are conceptual. In twentieth-century physics, particularly with the quantum revolution, this seemed to many to be the case. Along similar lines, the progress in molecular biology from 1945 to 1965 seemed nothing short of revolutionary in conceptual terms.[3] James Watson and Francis Crick's model of the double-helical structure of DNA, Crick's articulation of the "Central Dogma," and the cracking of the genetic code in the 1960s, begun by Marshall Nirenberg and Heinrich Matthei, provided a powerful new conceptual framework for understanding life at the molecular level. This period of intellectual ferment was not followed by stasis or senescence. Rather, in the early 1970s, the recombinant revolution changed how biologists visualize entities and organize their work. Even if these new techniques did not overturn the conceptual advances of the decades before, this did not mean that their consequences were any less sweeping. The development of tools for genetic engineering accelerated the acquisition of more-detailed knowledge in molecular biology and resulted in the "molecularization" of all of the life sciences, as biologists across the board adopted methods for cloning and manipulating genes—an enterprise spurred on by the publication by CSHL Press in 1982 of *Molecular Cloning: A Laboratory Manual.*[4] Molecular biology may have been dogmatic in the 1950s, but it became imperial in the 1980s. I would venture that this reflects a broader pattern in the experimental life sciences—that technological breakthroughs have played just as important a role as conceptual innovations in the development of modern biology.

The papers collected in this section represent some major contributions to this technological revolution of biology in the 1970s. The first of the set, Richard Roberts' 1976 paper on restriction enzymes, enumerates the restriction and modification enzymes discovered to date, listing also the microorganism sources and known restriction sites[5] (see also Roberts, p. 215). As Hans-Jörg Rheinberger has observed, the technical innovations in molecular biology in the 1970s and 1980s were "soft" technologies, in which biological agents became tools, rather than the "hard" technologies of biophysical instrumentation.[6] Biological entities that had previously been objects of scientific investigation themselves became the technical means for searching out other objects, above all genes. Restriction enzymes and ligases, the scissors and glue used to excise genes of interest from their chromosomal origins and place them in vectors for manipulation and scrutiny, exemplify what Rheinberger means by "soft" technology.

Roberts put restriction enzymes to use after arriving at Cold Spring Harbor in 1972 to examine the genome structure of adenovirus 2; research on the genes of this virus led his group to the stunning realization that it contained discontinuous genes[7] (see also Chow, *Tumor Viruses*, p. 105).

The third paper in this set introduced a related innovation, a new method for separating and identifying DNA fragments after cleavage with restriction enzymes[8] (see also Sambrook, p. 223). This was one of several breakthroughs that made the routine use of recombinant methods feasible. As Michel Morange has noted, "genetic engineering would not have developed so rapidly if new techniques, and the improvement of existing methods, had not simplified each step in the process."[9] Philip Sharp, Bill Sugden, and Joe Sambrook found that small amounts of cleaved DNA could be identified more easily in agarose gels that contained the dye ethidium bromide than by two methods in use at the time, velocity sedimentation and electrophoresis by polyacrylamide gels. Theirs was not simply a methods publication; the paper also demonstrated the existence of two new restriction activities in *Haemophilus parainfluenzae*. Even so, it was their use of agarose gels with ethidium bromide that caught on widely among molecular biologists, applied with great success in restriction analysis of animal viruses such as SV40. In turn, restriction mapping provided a crucial tool for the structural analysis of genes, ultimately laying the foundation for finer mapping in the form of base sequencing.

In 1976, Tom Maniatis, Sim Gek Kee, Argiris Efstratiadis, and Fotis Kafatos presented another innovative technique that, in turn, propelled the sequencing revolution. Their paper introduced a method for creating and amplifying complementary DNA, or cDNA, from messenger RNA encoding a protein of interest (in their case, hemoglobin)[10] (see also Maniatis, p. 227). The protein-to-gene approach to isolating DNA was not the only one developed—David Hogness and co-workers at Stanford fragmented the entire *Drosophila* genome to make a "library" with phage plasmids—but researchers soon realized that the cDNA, once created, could be used to clone the full chromosomal copy of the gene, with its regulatory sequences and introns. The development of these cloning techniques put individual eukaryotic genes within reach of biologists. These advances in recombinant methods, "technical wizardry," to use the words of Harrison Echols, were critical to the unfolding knowledge of eukaryotic gene expression, a major preoccupation of the 1970s and 1980s.[11]

The counterpart to being able to access and manipulate all of an organism's constituent genes is to be able to analyze the full range of functional proteins they produce under various conditions. James Garrels' 1989 paper in this set introduced a new system (QUEST) for not only visualizing, but also quantifying levels of cell proteins[12] (see also Garrels, p. 219). His system integrated two-dimensional gel electrophoresis (first introduced in 1975) with computer analysis with an aim to construct a protein database. Using this system, the resolution from radiolabeling enabled detection of 2000 proteins from an SV40-transformed cell line. Further interest in coupling experimental instrumentation with computers gained steam in the 1990s as part of the burgeoning interest in proteomics.[13] The trend was driven not only by basic science, but also by industrial biotechnology. Garrels himself represents this trend, having founded the company Proteome in 1995.

Why were these seminal technical contributions made at Cold Spring Harbor? Surely the emphasis there on tumor virology was critical, and this focus reflected Watson's imprint as Director. As Echols has noted, "Watson had exceptional skill in recognizing talented individuals, and he argued persuasively for the future importance of DNA animal viruses in research on eukaryotic molecular biology and cancer. He served as both a talent scout and prophet of a new era, a role very similar to that of Max Delbrück with the phage group 20 years earlier."[14] It is not that Watson sought to make Cold Spring Harbor a technology development center—to the contrary, Richards notes that in 1974 he "tried unsuccessfully to convince Jim Watson that Cold Spring Harbor should start a company to manufacture and sell restriction enzymes. He declined, thinking there was no money to be made."[15] But the new generation of "soft" technologies and their initial applications were developed by molecular biologists using the vanguard model organisms: bacteria and their phage and plasmids, cultured cell lines, and not least, tumor viruses. The linking of these technical developments to the War on Cancer had other unintended consequences. In particular, the initial recombinant experiments sparked concerns due to fears that tumor-inducing viruses spliced into bacteria, if released, might endanger public health.[16] Indeed, widespread fears concerning the use of recombinant DNA technologies are the reason Tom Maniatis came to CSH Laboratory to do his experiments; recombinant DNA procedures were at that time banned in Cambridge, Massachusetts[10] (see also Maniatis, p. 227). By the 1980s, the economic promise of genetic engineering and confidence in existing biological containment systems overshadowed these initial concerns. But whereas this dramatic political story of public concern and debates over regulation has received a great deal of attention, as have the first recombinant DNA experiments, the far-reaching effects of the mundane technical contributions recounted here remain less noticed. These four papers from Cold Spring Harbor Laboratory remind us that the development of workaday tools such as restriction enzymes and agarose electrophoresis, although rarely hailed as landmarks in the history of science, were indispensable to the development of molecular biology and its rapid spread to other areas of biomedical research and industry.

Notes and References

1. Stent G.S. 1968. That was the molecular biology that was. *Science* **160:** 390–395; quote on p. 390.

2. See Normal Science. Chapter 15 of Michel Morange, *A history of molecular biology*, trans. Matthew Cobb (1998 Harvard University Press, Cambridge, Massachusetts);

Thomas S. Kuhn, *The structure of scientific revolutions*, 2nd ed. (1962/1970 University of Chicago Press, Chicago, Illinois).

3. This sensibility is echoed in the title of one early synoptic account: Judson H.F. 1979. *The eighth day of creation: Makers of the revolution in biology*. Simon & Schuster, New York.

4. Maniatis T., Fritsch E.F., and Sambrook J. 1982. *Molecular cloning: A laboratory manual*. Cold Spring Harbor Laboratory Press, Cold Spring Harbor, New York.

5. Roberts R.J. 1978. Restriction and modification enzymes and their recognition sequences. *Gene* **4**: 183–194.

6. Rheinberger H.-J. 1995. Beyond nature and culture: A note on medicine in the age of molecular biology. *Science in Context* **8**: 249–263.

7. Chow L.T., Gelinas R.E., Broker T.R., and Roberts R.J. 1977. An amazing sequence arrangement at the 5′ ends of adenovirus 2 messenger RNA. *Cell* **12**: 1–8.

8. Sharp P.A., Sugden B., and Sambrook J. 1973. Detection of two restriction endonuclease activities in *Haemophilus parainfluenzae* using analytical agarose–ethidium bromide electrophoresis. *Biochemistry* **12**: 3055–3063.

9. Morange M., *History of molecular biology*, p. 197. (see ref. 2.)

10. Maniatis T., Kee S.G., Efstratiadis A., and Kafatos F.C. 1976. Amplification and characterization of a beta-globin gene synthesized in vitro. *Cell* **8**: 163–182.

11. Echols H. 2001. *Operators and promoters: The story of molecular biology and its creators*, p. 346. University of California Press, Berkeley.

12. Garrels J.I. 1989. The QUEST System for quantitative analysis of two-dimensional gels. *J. Biol. Chem.* **264**: 5269–5282.

13. Interest in protein expression databases also continued: By 1996, such databases had been constructed for plasma, rat and mouse liver, heart tissue, *E. coli*, and yeast. Anderson N.G. and Anderson N.L. 1996. Twenty years of two-dimensional electrophoresis: Past, present and future. *Electrophoresis* **17**: 443–453.

14. Echols H., *Operators and promoters*, p. 205. (see ref. 11.)

15. Richard J. Roberts, "Autobiography," http://nobelprize.org/nobel_prizes/medicine/laureates/1993/roberts-autobio.html

16. Among the many (and divergent) accounts of this controversy, see: Watson J.D. and Tooze J. 1981. *The DNA story: A documentary history of gene cloning*. W.H. Freeman, San Francisco, California; Krimsky S. 1982. *Genetic alchemy: The social history of the recombinant DNA controversy*. MIT Press, Cambridge, Massachusetts; and Wright S. 1994. *Molecular politics: Developing American and British regulatory policy for genetic engineering, 1972–1982*. University of Chicago Press, Illinois.

RICHARD J. ROBERTS

Roberts R.J. 1978. **Restriction and modification enzymes and their recognition sequences.** *Gene* **4**: 183–193. (Reprinted, with permission, from Elsevier ©1978.)

MOST BIOLOGISTS ARE NATURAL COLLECTORS of information. We like to know everything that has been previously done in our field so that we can know we are working at the cutting edge. In the 1970s, although a few biologists had discovered computers and were using them for collecting and archiving, most of us were managing with memory typewriters or primitive word processing systems. Working in the field of restriction endonucleases, I found that there was a great deal of interest in which enzymes were available for use and what their recognition sequences were, but relatively few publications describing them, and no single compilation other than the one in my own notebook. In 1974 for a meeting on restriction endonucleases in Ghent, Belgium, I showed a slide containing a list of the restriction enzymes known at that time. I also took a few extra print copies, which very quickly disappeared, and I left the meeting with a long list of names and addresses of people who wanted personal copies of the list. This led me to systematically maintaining this list, which led to the first formal publication in a peer-reviewed journal in *Gene* in 1978. This was the start of a lifelong mission for me to maintain a comprehensive database—REBASE—of information about restriction endonucleases that is now freely available through the Internet (http://rebase.neb.com/rebase/rebase.html).

Little did I realize in 1974 when I began collecting this information in a systematic fashion that I would be joining just a handful of other individuals who had seen the growing importance of collecting biological data in machine-readable form and that has led to the vast infrastructure of databases that are now key to the practice of biology. Today, we would make little progress without GenBank, PDB, Swiss-Prot, and the hundreds of other databases described in the annual Database Issues of Nucleic Acids Research.[1,2] Not only do these databases help us know what is known, but they also provide data upon which computation and data mining can take place.

RICH ROBERTS (*See biographical footnote on p. 51.*)

I see two main challenges going forward. One of these is to make sure that computer programs can easily traverse the interfaces into these databases, permitting simple connections to be made between data items in different databases. The second major challenge concerns the database that does not yet exist in a single form but which is finally showing signs of life. This is the database of the published literature, which is surely the greatest database of all. Most scientists of my age will recall their student days of poring through the stacks of journals in the basements of libraries trying to find key articles that were essential for our thesis or next paper. Adequately searching the print literature within a library is exceedingly difficult! However, searching the modern digitized text is easy, provided you have access to it.

For the last few years, I have been a strong advocate of open access publication, believing that all of science will best be served if both the current and the archival literature is completely and freely available to all scientists at a central location—preferably several—where it can be searched and properly mined for those nuggets of biological information that may have been forgotten because the experiments were carried out so long ago. Although I can to some extent understand the concerns of the commercial publishers who fear losing a profitable source of income, I have been appalled by the attitude of many academic publishers and scientific societies who seem to place financial returns above the health of the scientific enterprise. In my mind, it is inevitable that the literature will end up in a completely accessible digitized form in a database that will be seen to be the GenBank of the published literature. We should move fast to make it a reality.

Notes and References

1. Galperin M.Y. 2005. The Molecular Biology Database Collection. *Nucleic Acids Res.* **33**: D5–D24.
2. Galperin M.Y. 2006. The Molecular Biology Database Collection: 2006 update. *Nucleic Acids Res.* **34**(database issue): D3–D5.

Gene, 4 (1978) 183—193

183

© Elsevier/North-Holland Biomedical Press, Amsterdam — Printed in The Netherlands

Review

RESTRICTION AND MODIFICATION ENZYMES AND THEIR RECOGNITION SEQUENCES

RICHARD J. ROBERTS

Cold Spring Harbor Laboratory, P.O. Box 100, Cold Spring Harbor, NY 11724 (U.S.A.)

(Received September 7th, 1978)

(Accepted September 18th, 1978)

INTRODUCTION

During the last few years many bacterial strains have been examined for the presence of Type II restriction endonucleases and a large number of these enzymes have now been characterized. Much of the information available has never been formally published. While this reflects the lengthy time which can elapse between discovery and publication, increasingly it results from the fact that a newly discovered endonuclease is an isoschizomer of a more familiar one. Thus, unless the new source offers some advantage, there is a natural trend to avoid formal publication. To some extent, review articles [e.g. 86, 86a, 127] fill this gap; however, they quickly become outdated. The present compilation is an attempt to extend current awareness of the enzymes now available.

In forming this list, all endonucleases cleaving DNA at a specific sequence have been considered to be restriction enzymes, although in most cases, there is no direct genetic evidence for the presence of a restriction modification system. In addition to the examples listed, other strains are known to contain specific endonucleases. For example, many different *Staphylococcus aureus* isolates contain an isoschizomer of *Sau*3A [102], while a large-scale screening [63] of a wide range of gliding bacteria (orders: *Myxobacterales* and *Cytophagales*) has shown that 27 out of 120 strains examined contained restriction enzymes. These are not included in this list because, at the moment, they are insufficiently characterized.

Within the table, the source of each microorganism is given either as an individual or a National Culture Collection. If further information is required, it is available either in the first reference shown, which in each case refers to the purification procedure for the restriction enzyme, or from the individuals who provided their unpublished results. Where more than one reference appears, the second concerns the recognition sequence for the restriction enzyme; the third contains the purification procedure for the methylase, and the fourth describes the recognition sequence for the methylase. In some cases, two references appear in one of these categories when two independent groups have reached similar conclusions.

JAMES I. GARRELS

Garrels J.I. 1989. **The QUEST system for quantitative analysis of two-dimensional gels.** *J. Biol. Chem.* **264:** 5269–5282 (Reprinted, with permission, from American Society for Biochemistry and Molecular Biology, Inc. ©1989.)

L ONG BEFORE ANY "OMICS" TECHNOLOGIES EXISTED, Cold Spring Harbor Laboratory had the QUEST Protein Database Center. It consisted of a dedicated two-dimensional gel laboratory in the McClintock building, which was capable of running 20 large-format gels each day, and a state-of-the-art computer center located in the Carnegie Library (and later in the basement of Grace Auditorium) for analyzing the gel images. This paper, the first of a set of three contiguous papers, describes the quantitative gel analysis system. The following two papers,[1,2] written with Robert Franza, detailed our analysis of a set of normal and transformed cell lines (the REF52 family developed in the James Laboratory). At the time, we called our technology "Quantitative Two-dimensional Gel Electrophoresis of Proteins." Our work might have received more attention if we had thought to call it "Proteomics."

The origins of the QUEST lab at Cold Spring Harbor go back to 1977 when Jim Watson visited the University of California at San Diego and the Salk Institute, where I was doing graduate research with David Schubert, a cell culture expert who had a marvelous collection of nerve and muscle cell lines. I had recently published a two-dimensional gel paper with Wade Gibson of Salk's Tumor Virus Laboratory, in which we identified two forms of actin in nonmuscle cells and three forms in muscle cells. Contractile proteins of the cytoskeleton were just getting hot, and because of Jim's special interest in the Cold Spring Harbor Cell Biology group, he paid me an unannounced visit. I showed my work and used the opportunity to ask if I could get into the 1977 Symposium on Chromatin. No thought of working at the Lab occurred

JIM GARRELS, a native Iowan, holds a B.S. in biology and physics from the California Institute of Technology and a Ph.D. from the University of California at San Diego. He worked at Cold Spring Harbor for 17 years (1978–1995) before founding Proteome, Inc., in Beverly, Massachusetts. Proteome was sold to Incyte Genomics in 2000, and Jim left the company a year later. In 2005, Jim and his wife Joan Brooks cofounded a new company, Garbrook Associates, to build knowledge resources for the renewable energy field. jgarrels@garbrook.com

Jim Garrels, Quest Computer Lab, 1978. *(Courtesy of Cold Spring Harbor Laboratory Archives.)*

to me until I saw Jim again at the meeting. He introduced me to the Cell Biology Group, which included Keith Burridge, Lan Bo Chen, and Guenter Albrecht-Buehler, as if I was already a new member of the group. I was in a bind because I had already accepted a postdoc position with Bruce Alberts at the University of California at San Francisco, but a call from Jim quickly made things right with Bruce.

Coming to Cold Spring Harbor in 1978, I had great support from Jim Watson. He added an extension onto the McClintock Laboratory to house the two-dimensional (2D) gel lab, and he guaranteed funds for my computer facility, even though computers were not yet common tools for biology. A grant from the National Science Foundation (NSF) was secured to fund the facility, and soon, a PDP-11/60 computer (the first computer at Cold Spring Harbor Laboratory) and an 88-megabyte disk drive as big as a dishwasher were installed through a window into a second-floor room of Carnegie. The gel lab was custom-designed with large sink/benches for the wide 2D slab gel units, and a semiautomated dipping system was installed in a fume hood to treat the gels with chemicals for fluorographic detection of the radiolabeled proteins. The lab ran gels for projects by various groups at Cold Spring Harbor and for many collaborators in other institutions. Bob Franza (see p.

199) joined me in the early 1980s to help build a 2D gel database for cancer cells. The REF52 database contained quantitative data for 1600 proteins from each of 79 cell cultures. These cultures represented normal cells, cells transformed with DNA tumor viruses, and cells transformed with retrovirus. These studies revealed tightly regulated groups of proliferation-sensitive proteins (such as proliferating cell nuclear antigen, PCNA) that had lost regulation in the cancer cells.

The work was exciting, well-funded, and well-received; however, there were many in those years who simply saw no point in looking at thousands of proteins (or genes) at once. The system did have limitations, and one of the largest was our inability to identify many of the proteins we had detected. Rapid identification of proteins would have to wait for the availability of gene sequences, and the first genome project (for yeast) was just getting under way in 1989.

After publication of these papers, I turned to the study of yeast because of the emerging genome sequence and because of the large number of interesting mutants available for study. This period led to interesting and fruitful collaborations[3] with Cal McLaughlin of the University of California, Irvine, Jonathan Warner of Albert Einstein College of Medicine, and Bruce Futcher of Cold Spring Harbor. We succeeded in the identification of many proteins and in the characterization of mutants involved in heat shock, protein modification, and RNA splicing.

After leaving Cold Spring Harbor to form a new company in 1995, I continued my interest in protein databases, although I eventually left the 2D gel field to concentrate on building protein databases by curation of genomic sequences and the published literature. The knowledge of proteins from proteomic and genomic studies continues to grow at an ever-increasing rate, far more than could ever have been imagined when the QUEST papers were published in 1989.

Notes and References

1. Garrels J.I. and Franza B.R. Jr. 1989. The REF52 protein database. Methods of database construction and analysis using the QUEST system and characterizations of protein patterns from proliferating and quiescent REF52 cells. *J. Biol. Chem.* **264:** 5283–5298.

2. Garrels J.I. and Franza B.R. Jr. 1989. Transformation-sensitive and growth-related changes of protein synthesis in REF52 cells. A two-dimensional gel analysis of SV40-, adenovirus-, and Kirsten murine sarcoma virus-transformed rat cells using the REF52 protein database. *J. Biol. Chem.* **264:** 5299–5312.

3. See, for example: Garrels J.I., McLaughlin C.S., Warner J.R., Futcher B., Latter G.I., Kobayashi R., Schwender B., Volpe T., Anderson D.S., Mesquita-Fuentes R., and Payne W.E. 1987. Proteome studies of *Saccharomyces cerevisiae*: Identification and characterization of abundant proteins. *Electrophoresis* **18:** 1347–1360.

THE JOURNAL OF BIOLOGICAL CHEMISTRY
© 1989 by The American Society for Biochemistry and Molecular Biology, Inc.

Vol. 264, No. 9, Issue of March 25, pp. 5269–5282, 1989
Printed in U.S.A.

The QUEST System for Quantitative Analysis of Two-dimensional Gels*

(Received for publication, October 12, 1988)

James I. Garrels

From the Cold Spring Harbor Laboratory, Cold Spring Harbor, New York 11724

The strategies and methods used by the QUEST system for two-dimensional gel analysis are described, and the performance of the system is evaluated. Radiolabeled proteins, resolved on two-dimensional gels and detected using calibrated exposures to film, are quantified in units of disintegrations per minute or as a fraction of the total protein radioactivity applied to the gel. Spot quantitation and resolution of overlapping spots is performed by two-dimensional gaussian fitting. Pattern matching is carried out for groups of gels called matchsets, and within each matchset every gel is matched to every other gel. During the matching process, spots are automatically added to each pattern at positions where unmatched spots were detected in other patterns. This results in enhanced accuracy for both spot detection and for matching. The spot fitting procedure is repeated after matching. Tests show that up to 97% of spots in each pattern can be matched and that fewer than 1% of the spots are matched inconsistently.

Approximately 2000 proteins are detected from typical gels. Of these 1600 are high quality spots. Tests to measure the coefficient of variation of spot quantitation *versus* spot quality show that the average coefficient of variation for high quality spots is 21%. The intensities of the detected proteins range from 4 to 20,000 ppm of total protein synthesis. The QUEST analysis system has been used to build a quantitative database for the proteins of normal and transformed REF52 cells, as presented in the accompanying reports (Garrels, J., and Franza, B. R., Jr. (1989) *J. Biol. Chem.* 264, 5283–5298, 5299–5312).

Two-dimensional gel electrophoresis is a method for the analytical separation of large numbers of proteins from solubilized biological samples (1). The resolving power of the technique, which combines isoelectric focusing in the first dimension and sodium dodecyl sulfate electrophoresis in the second dimension, is high because it separates proteins by two independent physical properties, charge and molecular mass. Protein patterns containing more than 2000 spots have been obtained.

Given the ability to resolve so many proteins from cells without prior fractionation, many researchers have hoped that the technique could be used for quantitative and systematic

* This work was supported by Grant P41 RR02188 from the National Institutes of Health Division of Research Resources, by a grant from the Cancer Research Institute, by Grant 321100 from the New York State Science and Technology Foundation, and by National Institutes of Health Grant CA 13106. The costs of publication of this article were defrayed in part by the payment of page charges. This article must therefore be hereby marked "*advertisement*" in accordance with 18 U.S.C. Section 1734 solely to indicate this fact.

studies of cellular regulation. Using computers to quantify and compare the protein patterns, databases could be constructed to accumulate and communicate detailed information about each of the detected proteins. Protein databases would give meaning to the complex patterns by storing for each detected protein such information as its name, its subcellular localization, its known modifications, and the regulation of its synthesis and turnover under many experimental conditions.

Numerous problems have been faced by those attempting to exploit the full power of two-dimensional gels. First, the technique is difficult and lacks standardization. Many groups that use the technique extensively have modified the original procedure for improved resolution, reproducibility, and ease of use (2–7); however, comparisons of protein patterns between laboratories are still difficult. Second, quantitation from two-dimensional gels requires special calibration procedures. Some groups have developed procedures to calibrate the films used for autoradiography or fluorography (5, 8, 9), and others have used quantitative staining procedures (10). Third, algorithms have had to be developed to detect and to quantify the spots from two-dimensional gel images. One method is to define each spot by its boundary contour (9), and another is to model each spot by gaussian curve fitting (11, 12). Fourth, pattern matching is difficult because the electrophoretic patterns from different gels are not completely superimposable and because spots in overlapping clusters are not always resolved identically in every gel. Several matching algorithms using neighborhood information have been developed (8, 13, 14), and methods to resolve the ambiguities of matching have been explored (8, 14, 15). Finally, the design of the protein databases themselves presents a formidable challenge. Various database structures have been proposed by the Anderson group (16, 17), by Tarroux (18, 19), by Lipkin and Lemkin (20, 21), and in our own previous work (8); however, the actual production of quantitative computeraccessible protein databases has been very limited.

The QUEST system is an integrated system of electrophoresis and computer analysis designed for the construction of protein databases. The laboratory techniques and equipment used for the routine production of high quality gels have been reported elsewhere (2, 5). In this report, the methods for analysis of two-dimensional gel patterns are presented. In the following report (22), the methods of database construction and database analysis are presented; and, in the third paper of this series (23), the complete system is applied to the study of growth regulation and viral transformation in REF52 cells.

Many of the algorithms and strategies for two-dimensional gel analysis presented here differ considerably from those presented in a previous version of the QUEST system (8). The present software is completely written in the C language for use on modern UNIX-based workstations. Major new features include 1) the ability to combine images from sepa-

JOSEPH F. SAMBROOK

Sharp P.A., Sugden B., and Sambrook J. 1973. **Detection of two restriction endonuclease activities in** *Haemophilus parainfluenzae* **using analytical agarose–ethidium bromide electrophoresis.** *Biochemistry* **12:** 3055–3063. (Reprinted, with permission, from the American Chemical Society ©1973.)

I REMEMBER A CONVERSATION IN LONDON with Renato Dulbecco in late 1971, around the time that he moved from the Salk Institute to the Imperial Cancer Research Fund. We were both excited about the work of Kathleen Danna and Dan Nathans that was to be published in December of that year, showing that a restriction enzyme isolated from *Haemophilus influenzae* cleaved SV40 DNA into 11 specific fragments.[1] It was obvious to us, as to many others, that restriction enzymes were about to make accessible much that had previously been beyond our reach.

At Cold Spring Harbor, our need for restriction enzymes was immediate. We were completing work on the assignment of SV40 mRNAs to the early and late strands of the viral DNA, and the next step was to map the locations of the early and late transcripts on the viral genome. So we began purifying restriction enzymes with the general aim of establishing the technology at Cold Spring Harbor and the specific hope of finding an enzyme that would cut SV40 DNA at a limited number of well-spaced sites that could be easily mapped and used to chart the transcription of the virus.

Why we lit upon *H. parainfluenzae* I no longer remember. Perhaps for no other reason than Jane Setlow at nearby Brookhaven National Laboratory was holding a stock of the bacteria. Anyway, it was certainly a fortuitous choice. Using the ancient fermenter in the basement of Demerec Laboratory, Bill Sugden, a graduate student who was just finishing his Ph.D. at Cold Spring Harbor, and I grew 10-liter cultures and began experimenting with different ways to break open the bacteria and with different resins to fractionate proteins. To our pleasure, we found that *H. parainfluenzae* contained two restriction enzymes that cleaved SV40 DNA three times and once, respectively. The resulting DNA fragments turned out to be of a size and location ideal for mapping of SV40 mRNAs by hybridization and, within a few months, had been used successfully to produce a low-level transcription map of the viral genome.[2]

JOE SAMBROOK (*See biographical footnote on p. 37.*)

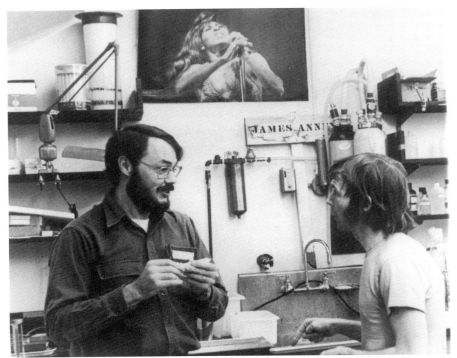

Phil Sharp, Tina Turner, and Joe Sambrook, circa 1973. *(Courtesy of Cold Spring Harbor Laboratory Archives.)*

We wanted to search other strains of bacteria for useful enzymes, but assaying large numbers of column fractions for restriction activity was a slow and laborious process. Labeling DNAs in vitro still lay some years in the future, so the assay required viral DNA substrates labeled in vivo to high specific activity with ^{32}P and autoradiography of gels. The weekly grind of labeling and purifying SV40 DNA, coupled with the delays in enzyme purification imposed by exposure of autoradiographs, quickly became tiresome.

The notion of using ethidium bromide to stain unlabeled DNA in gels occurred to two groups simultaneously. The procedure used by Aaij and Borst[3] involved immersing the gel in concentrated dye solution followed by a lengthy destaining process to reduce the background levels of fluorescence. We took a different tack, using low concentrations of ethidium bromide that could be either incorporated into the gel or used for staining after electrophoresis was completed. This technique, both rapid and sensitive, is still widely used in an essentially unaltered form.

Other people working in James Laboratory began to use our assay both to screen other bacterial species for restriction activities and to prepare enzymes for general use in the laboratory. After a while, we set up a scheme where Ronni Greene, a research assistant, would produce the set of restriction enzymes needed for work on SV40 and adenoviruses. A few months later, Rich Roberts, who had recently arrived from Harvard and who was casting around for something to do at Cold Spring Harbor, appeared in James Laboratory and rather nervously asked whether we would show him how to purify and assay restriction enzymes. The people in James were happy to help and after working with them for a week or so, Rich began his own search for new restriction enzymes.

For several years afterward, both our laboratory and Rich's continued to make enzymes for use within Cold Spring Harbor and beyond, a task that became increasingly onerous as the number of useful enzymes grew. By the mid 1970s, restriction enzymes were an attractive commercial proposition. Rich became involved in New England BioLabs while he and Phyllis Myers, his research assistant at Cold Spring Harbor, continued to search for new enzymes. James Laboratory, with some reluctance, discovered it could buy restriction enzymes more cheaply than it could make them; Ronni Greene resumed a life at the research bench and we all turned our hands to other projects.

Notes and References

1. Danna K.J. and Nathans D. 1971. Specific cleavage of simian virus 40 DNA by restriction endonuclease of *Hemophilus influenzae. Proc. Natl. Acad. Sci.* **68:** 2913–2917.

2. Sambrook J., Sugden B., Keller W., and Sharp P.A. 1973. Transcription of simian virus 40. III. Mapping of "early" and "late" species of RNA. *Proc. Natl. Acad. Sci.* **70:** 3711–3715.

3. Aaif C. and Borst P. 1972. The gel electrophoresis of DNA. *Biochim. Biophys. Acta.* **269(2):** 192–200.

Detection of Two Restriction Endonuclease Activities in *Haemophilus parainfluenzae* Using Analytical Agarose–Ethidium Bromide Electrophoresis†

Phillip A. Sharp,* Bill Sugden, and Joe Sambrook

ABSTRACT: A rapid assay for restriction enzymes has been developed using electrophoresis of DNA through 1.4% agarose gels in the presence of 0.5 μg/ml of ethidium bromide. The method eliminates lengthy staining and destaining procedures and resolves species of DNA which are less than 7×10^6 daltons. As little as 0.05 μg of DNA can easily be detected by direct examination of the gels in ultraviolet light. Using this technique, we have identified two different restricting activities in extracts of *Haemophilus parainfluenzae*. The two activities have different chromatographic properties on phosphocellulose and Bio-Gel A-0.5m, and they attack SV40 DNA at different sites. One activity (*Hpa* II) cleaves SV40 DNA at different sites. One activity (*Hpa* II) cleaves SV40 DNA at a single position situated 0.38 fractional genome length from the insertion point of SV40 sequences into the adenovirus SV40 hybrid Ad2++ND$_1$. The other activity (*Hpa* I) cleaves SV40 DNA at three sites which appear to coincide with 3 of the 11 cleavage points attacked by a restriction system isolated from *H. influenzae* strain *Rd*.

Restriction enzymes have been isolated from a variety of strains of *Escherichia coli* and from various species of *Haemophilus* (see review by Meselson *et al.*, 1972; Yoshimori, 1971; Gromkova and Goodgal, 1972; Sack and Nathans, 1973). All of them are highly specific endonucleases which produce double-strand cleavages of native unmodified DNA. At least three of these enzymes (*Haemophilus influenzae* (Kelly and Smith, 1970), *E. coli* R·R$_I$ (Hedgpeth *et al.*, 1972; Mertz and Davis, 1972), and *E. coli* R·R$_{II}$) (H. W. Boyer, personal communication) attack base sequences that possess twofold rotational symmetry about an axis perpendicular to the axis of the DNA duplex; in other words, palindromic sequences of the type

$$\begin{array}{cccccc} A & B & C & C' & B' & A' \\ A' & B' & C' & C & B & A \end{array}$$

Because different enzymes attack different palindromes, each of them generates a characteristic set of cleavage products when reacted with DNA. For any particular enzyme the number of fragments obtained is a measure of the number of palindromic sites in the DNA specific to the enzyme, and the size of the fragments reflects the distribution of the sites along the DNA.

The two principal methods which have been used to analyze the fragments of DNA produced by restriction enzymes are velocity sedimentation and electrophoresis through poly-

† From Cold Spring Harbor Laboratory, Cold Spring Harbor, New York 11724. *Received April 13, 1973.* Supported by Grant CA11432 from the National Cancer Institute, U. S. Public Health Service.

Maniatis T., Kee S.G., Efstratiadis A., and Kafatos F.C. 1976. **Amplification and characterization of a β-globin gene synthesized in vitro.** *Cell* **8**: 163–182. (Reprinted, with permission, from Elsevier ©1976.)

THIS PAPER WAS PUBLISHED WHILE I WAS WORKING at Cold Spring Harbor Laboratory with a joint appointment as an Assistant Professor at Harvard.[1] The report describes the first cloning and characterization of a nearly full-length double-stranded synthetic cDNA containing the entire protein-coding sequence of a eukaryotic gene. The double-stranded cDNA in this publication and in an earlier paper[2] was referred to as a "gene," because the work was done prior to the discovery of RNA splicing, at a time when a gene was considered to be the DNA sequence encoding an mRNA, on the basis of what we knew from bacteria. The project began while I was still a postdoctoral fellow in Mark Ptashne's laboratory at Harvard and continued when I became a Senior Scientific Staff member at Cold Spring Harbor. As a postdoc, I characterized individual repressor-binding sites in the two operators of bacteriophage λ DNA with the goal of understanding the genetic switch from lysogenic to lytic gene expression. I had discovered that the λ operators contain multiple repressor-binding sites and was attempting to determine the size of the DNA sequences protected from digestion with DNase I by different concentrations of λ repressor. To accomplish this, I collaborated with Hans van deSande, then a postdoc in Gobin Khorana's laboratory at the Massachusetts Institute of Technology, who provided synthetic DNA sequences of known size that could be used as size markers. With these markers, it was possible to show that small DNA fragments could be separated according to size on denaturing polyacrylamide gels

TOM MANIATIS trained in molecular biology at Vanderbilt University and moved in 1971 as post-doctoral fellow to Mark Ptashne's laboratory at Harvard. Ptashne was continuing to anatomize the λ operator and Maniatis joined him in using restriction enzymes to dissect the functional components of the operator. Maniatis by heroic means determined the sequence of 33 (out of 100) bases of the leftward operator and no fewer than 74 base pairs of the rightward operator—all before the development of the chemical and dideoxy sequencing methods (1975). At about the same time, recombinant DNA methods, in particular cDNA cloning, made it possible to do molecular analysis of eukaryotic genes, and Maniatis and his colleagues Argiris Efstratiadis and Fotis Kafatos, as described in this chapter, were among the first to take up the challenge. maniatis@mcb.harvard.edu

Tom Maniatis, 1977. *(Courtesy of Cold Spring Harbor Laboratory Archives.)*

(containing urea or formamide), that the electrophoretic mobilities of the single-stranded DNA fragments are a linear function of the logarithm of their chain lengths, and that it was possible to separate DNA fragments that differ by only a single nucleotide.[3] Thus, with appropriate size markers, the size of small DNA fragments could be determined at a single-nucleotide resolution. These observations not only made possible the precise identification of individual λ repressor-binding sites, but also provided the basis for using denaturing polyacrylamide gel electrophoresis to determine the sequence of DNA.[4]

In the summer of 1974, while still working on bacteriophage λ, I met Arg Efstratiadis, a graduate student in Fotis Kafatos' laboratory in the Biolabs at Harvard. Although Arg had earlier received his M.D. degree in Greece, he wanted to focus mainly on research. While working on a doctoral thesis in lipid biochemistry, he attended the famous summer course in molecular biology on the island of Spetsai and this prompted his switch from medicine to molecular biology. In fact, he became so excited with the new science by reading Watson's *Molecular Biology of the Gene* (which he later translated into Greek) that he decided to restart his career by pursuing a Ph.D. at Harvard. When we first met, Arg was attempting to produce full-length cDNA copies of eukaryotic mRNAs, but he was unable to accurately determine the size of the cDNA products. His goal was to generate hybridization probes for studying mRNA developmental profiles in particular tissues of silk moths, which was the primary focus in the Kafatos lab. Fotis was among the first to apply biochemical approaches to insect development and was catapulted into a faculty position and early tenure at Harvard through his discovery and biochemical characterization of the enzyme cocoonase. Because of the paucity of insect material, Arg was optimizing the conditions of cDNA synthesis by using rabbit globin mRNAs.

At that time, I was the only person at Harvard using restriction enzymes, which I had initially obtained from Rich Roberts at Cold Spring Harbor Laboratory, but was by then purifying a few on my own. Arg stopped by Ptashne's lab to ask my technician Andrea Jeffrey for restriction fragments of λ DNA that he wanted to use as size markers in alkaline sucrose gradients. I happened to overhear their conversation and interrupted to suggest that he instead use formamide gels for accurate sizing of single-stranded DNA. Subsequent technical and scientific conversations evolved into an intense and productive multiyear collaboration and lifelong friendship.

As the work progressed, Arg, Fotis, and I became very excited with the prospect of generating and cloning double-stranded cDNA using methods that had just been developed at Stanford and the University of California, San Francisco, by Lobban and Kaiser[5] and by Cohen and Boyer.[6] Arg and Fotis were developmental biologists interested in understanding regulatory mechanisms underlying cellular differentiation, whereas I was interested in extending mechanistic studies of gene regulation in λ to eukaryotes. At that time, the primary approach to studies of eukaryotic gene expression was based on nucleic acid hybridization kinetics pioneered by Roy Britten and Eric Davidson.[7] Like others in the field, we thought that cDNA cloning would make it possible to study directly individual genes in eukaryotic cells through the use of gene-specific hybridization probes.

Initially, Arg optimized conditions for the synthesis and characterization of full-length single-stranded cDNA molecules from globin mRNAs using reverse transcriptase.[8] cDNA synthesis was widely used at the time to generate probes, but the reaction had not been optimized to synthesize and characterize full-length cDNAs. In addition, the sizes of reverse transcripts were only crudely determined by velocity sedimentation on sucrose density gradients. The use of denaturing polyacrylamide gel electrophoresis provided a fast accurate tool for cDNA characterization. With high concentrations of deoxynucleotide triphosphates and optimized reaction conditions, full-length globin cDNAs could be generated. Interestingly, S1 nuclease resistance results provided a clue (also noticed independently by others)[9] that a 3'-terminal hairpin structure was generated during reverse transcription of mRNA (S1 nuclease is an enzyme that cleaves single-stranded, but not double-stranded, DNA). We attempted to use the 3' end of the first cDNA strand to generate the second complementary strand, but the efficiency was very low with reverse transcriptase. We therefore tried to use *Escherichia coli* DNA polymerase I for second-strand synthesis.[2] Under the appropriate conditions, a full-length second cDNA strand could be generated, evidenced by the observation that the two cDNA strands were covalently joined and were twice the size of the first cDNA strand on denaturing polyacrylamide gels. We also found that it was possible to cleave the hairpin with S1 nuclease. Thus, it was possible to generate nearly full-length double-stranded cDNA, which was characterized using a set of three restriction enzymes.[1] In the Discussion section of the paper, we pointed out that, in principle, the newly developed recombinant DNA methods could be used to clone double-stranded cDNA and thus obtain homogeneous populations from a complex mixture of mRNAs.

An indication of trouble on the horizon was a section in the Discussion subtitled "Safety Considerations," which began "The availability of pure defined genes should enhance the acceptability of work with eukaryotic DNA." This seems like a strange statement now, but at that time, it reflected growing opposition to the use of recombinant DNA methods at Harvard, led by faculty members George Wald, Ruth Hubbard, and Ursula Goodenough.

As we began preparations to attempt cDNA cloning, the opposition to recombinant DNA research exploded into a public debate in the Cambridge City Council, largely due to conversations between George Wald and Al Vellucci, the Harvard-baiting Mayor of Cambridge. This led to City Council hearings and a moratorium on DNA cloning within the Cambridge city limits. We considered doing our experiments at the Harvard Medical School across the river in Boston, where there was no moratorium, but Charlie Thomas[10] made that difficult by his defiance of the restrictions on recombinant DNA research at the Medical School. This conflict ultimately led Charlie to move to California. Fortunately, Jim Watson had heard of the problems in Cambridge and offered to provide a lab to carry out the cDNA cloning at Cold Spring Harbor. This part of the history is discussed in more detail in *Inspiring Science: Jim Watson and the Age of DNA.*[11]

During the turmoil, I was beginning to establish my lab at Harvard, and among the first graduate students to join was Gek Kee Sim, who had recently graduated from Brandeis University. Gek Kee was bright, hard working, and exceptionally talented in the lab. She joined the cDNA cloning effort after I moved to Cold Spring Harbor, and she and I worked closely together to establish enzymatic procedures necessary for cloning and colony hybridization methods to identify the clones. We worked in a small kitchen area in Demerec, down the hall from Rich Roberts' lab. We were knocking elbows with other members of my budding lab, including David Goldberg, Elizabeth Lacy, and Lydia Villa Kamoroff (a postdoc from Fotis' lab) crammed together in a small room that formerly served as a kitchen for media preparation. We chose the dA:dT-cloning procedure developed at Stanford by Peter Lobban (a student in Dale Kaiser's lab) and the plasmid pMB9, a smaller derivative of the Cohen/Boyer plasmid pSC101 derived by Herb Boyer. This cloning method had been used earlier by Peter Wensink in Dave Hogness' lab to clone fragments of *Drosophila* genomic DNA.[12]

During this era, which predated reagent kits by many years, we had to purify many of the enzymes we used, such as T4 polynucleotide kinase and λ exonuclease. In addition, we begged other enzymes such as *E. coli* DNA polymerase, *E. coli* RNA polymerase, and RNase H from colleagues.

Although we were able to obtain a sample of terminal transferase, we made a large preparation of our own enzyme from calf thymus obtained from a Brooklyn slaughterhouse. Gek Kee and I made a huge mess homogenizing the tissue in the kitchen of Blackford Hall. Characterization of the cDNA clones required multiple restriction enzymes, which were generously provided by Rich Roberts and Phyllis Myers just down the hall in Demerec. Arg produced the double-stranded DNA in Cambridge and mailed it to Cold Spring Harbor. We treated the cDNA with λ exonuclease to prevent nonuniform dA:dT tailing and then with terminal transferase to add the dT tails. The plasmid was cleaved with *Eco*RI and treated with λ exonuclease, and a tail of dA residues was added to the 3′ terminus. To show that the mRNA had been copied into DNA without error and faithfully amplified in bacteria, the cDNA inserts obtained were thoroughly characterized. In fact, this is one of the reasons we chose to clone the rabbit β-globin mRNA. It was not only a highly abundant mRNA in red blood cells, but also the most thoroughly characterized eukaryotic mRNA at the time. Thus, we created the first detailed restriction map of a double-stranded cDNA using Robert's restriction enzymes. We then matched the map with an RNA sequence predicted from the complete amino acid sequence of the protein and from the known sequences of some small regions of β-globin mRNA.

Although we used the colony hybridization method of Grunstein and Hogness[13] to identify the bacterial colonies carrying the globin cDNA, this was hardly necessary; 80% of the clones were positive. Thus, we could have examined each clone directly using restriction enzymes. In any case, we went on to show that the restriction map of the cloned DNA was identical to the synthetic double-stranded cDNA, indicating that no rearrangements had occurred, and at least at the level of restriction sites, there were no mutations. We also determined the number of dA-dT residues used to join the cDNA to the plasmid and concluded that the cDNA was lacking only about ten nucleotides at the 5′ end, probably a consequence of the S1 nuclease digestion. In a subsequent paper, we sequenced the entire cDNA insert, which confirmed the faithful propagation of the cDNA in *E. coli*, and provided the first complete nucleotide sequence of an mRNA.[14] Subsequently, Gek Kee went on to construct a cDNA library and demonstrate the feasibility of using cDNA libraries of complex mRNA populations to obtain individual mRNAs.[15]

Our work, along with that of two other groups—Rougeon, Kourilsky, and Mach at Geneva and Paris,[16] and Terry Rabbitts at the MRC labs in Cambridge England[17]—established a powerful new approach to the study of molecular and cellular biology. cDNA-cloning methods have provided gene-specific probes for unraveling the complexity of eukaryotic gene regulation, they have led to major advances in our understanding of the structure and function of eukaryotic proteins, and they have played a key role in the large-scale production of proteins for treating a variety of human diseases. The latter two applications require the cloning of full-length cDNAs, which was the unique advance provided by our contributions. In our naiveté and that of Cold Spring Harbor, we made no attempt to patent the method, which was widely used in the biotechnology industry to produce protein-based drugs.

The relevance of this history to this book is that it provides yet another example of Jim's foresight and ability to make things happen. Because Jim recognized the potential of cDNA cloning and was unhappy with the politics in Cambridge, he made it possible to continue studies that could not be done in Harvard.

Notes and References

1. Maniatis T., Kee S.G., Efstratiadis A., and Kafatos F.C. 1976. Amplification and characterization of a β-globin gene synthesized in vitro. *Cell* **8:** 163–192.

2. Efstratiadis A., Kafatos F.C., Maxam A.M., and Maniatis T. 1976. Enzymatic in vitro synthesis of globin genes. *Cell* **7:** 279–288.

3. Maniatis T., Jeffrey A., and van deSande H. 1975. Chain length determination of small double- and single-stranded DNA molecules by polyacrylamide gel electrophoresis. *Biochemistry* **14:** 3787–3794.

4. Maxam A.M. and Gilbert W. 1977. A new method for sequencing DNA. *Proc. Natl. Acad. Sci.* **74:** 560–564.

   Sanger F. and Coulson A.R. 1975. A rapid method for determining sequences in DNA by primed synthesis with DNA polymerase *J. Mol. Biol.* **94:** 441–448.

5. Lobban P.E. and Kaiser A.D. 1973. Enzymatic end-to-end joining of DNA molecules. *J. Mol. Biol.* **78:** 453–471.

6. Cohen S.N., Chang A.C., Boyer H.W., and Helling R.B. 1973. Construction of biologically functional bacterial plasmids in vitro. *Proc. Natl. Acad. Sci.* **70:** 3240–3244.

   Morrow J.F., Cohen S.N., Chang A.C., Goodman H.M., and Helling R.B. 1974. Replication and transcription of eukaryotic DNA in *Escherichia coli*. *Proc. Natl. Acad. Sci.* **71:** 1743–1747.

7. Davidson E.H. and Britten R.J. 1973. Organization, transcription, and regulation in the animal genome. *Q. Rev. Biol.* **48:** 565–613.

8. Efstratiadis A., Maniatis T., Kafatos F.C., Jeffrey A., and Vournakis J.N. 1975. Full length and discrete partial reverse transcripts of globin and chorion mRNAs. *Cell* **4:** 367–378.

9. Salser W.A. 1974. DNA sequencing techniques. *Annu. Rev. Biochem.* **43:** 923–965.

10. Tom got into a bureaucratic tangle with the National Institutes of Health and Harvard concerning whether his laboratory at the Medical School was or was not certified to carry out recombinant DNA experiments involving SV40. Faced with administrative rigidity in dealing with risks that he believed to be entirely imaginary, Charlie left Harvard in a huff for the sunnier climate of Southern California.

11. Maniatis T. 2003. Cold Spring Harbor and recombinant DNA. In *Inspiring Science: Jim Watson and the age of DNA*. (ed. J. Inglis et al.), pp. 321–327. Cold Spring Harbor Laboratory Press, Cold Spring Harbor, New York.

12. Wensink P.C., Finnegan D.J., Donelson J.E., and Hogness D.S. 1974. A system for mapping DNA sequences in the chromosomes of *Drosophila melanogaster*. *Cell* **3:** 315–325.

13. Grunstein M. and Hogness D.S. 1975. Colony hybridization: A method for the isolation of cloned DNAs that contain a specific gene. *Proc. Natl. Acad. Sci.* **72:** 3961–3965.

14. Efstratiadis A., Kafatos F.C., and Maniatis T. 1977. The primary structure of rabbit beta-globin mRNA as determined from cloned DNA. *Cell* **10:** 571–585.

15. Sim G.K., Efstratiadis A., Jones C.W., Kafatos F.C., Koehler M., Kronenberg H.M., Maniatis T., Regier J.C., Roberts B.F., and Rosenthal N. 1978. Studies on the structure of genes expressed during development. *Cold Spring Harbor Symp. Quant. Biol.* **42:** (part 2) 933–945.

    Sim G.K., Kafatos F.C., Jones C.W., Koehler M.D., Efstratiadis A., and Maniatis T. 1979. Use of a cDNA library for studies on evolution and developmental expression of the chorion multigene families. *Cell* **18:** 1303–1316.

16. Rougeon F., Kourilsky P., and Mach B. 1975. Insertion of a rabbit beta-globin gene sequence into an *E. coli* plasmid. *Nucleic Acids Res.* **2:** 2365–2378.

17. Rabbitts T.H. 1976. Bacterial cloning of plasmids carrying copies of rabbit globin messenger RNA. *Nature* **260:** 221–225.

Cell, Vol. 8, 163–182, June 1976, Copyright © 1976 by MIT

Amplification and Characterization of a β-Globin Gene Synthesized in Vitro

Tom Maniatis,*† **Sim Gek Kee,***†
Argiris Efstratiadis,† **and Fotis C. Kafatos**†
*Cold Spring Harbor Laboratory
Cold Spring Harbor, New York 11724

†Harvard University Biological Laboratories
Cambridge, Massachusetts 02138

Summary

Full-length, double-stranded globin DNA was synthesized in vitro starting from rabbit globin mRNA. Several restriction endonuclease cleavage sites with known recognition sequences were mapped on this DNA as a means of assessing the accuracy of in vitro synthesis. By comparing this map with the nucleotide sequences known or predicted from the amino acid sequences of α- and β-chain rabbit hemoglobin, it was possible to show that the synthetic globin DNA is a faithful copy of β-globin mRNA.

Amplification of the synthetic globin DNA was achieved by inserting the molecule into the plasmid PMB9 using the poly(dA)•(dT) joining procedure, and transforming E. coli with the hybrid DNA. Transformants carrying β-globin DNA were identified by colony hybridization using purified ¹²⁵I–β–mRNA probe. Comparison of the restriction maps of the synthetic and inserted globin DNAs showed that the entire synthetic globin DNA molecule was amplified without sequence rearrangements.

Both the synthetic and the cloned DNA include the entire coding sequence of the β-globin gene plus a substantial portion of the untranslated regions flanking the structural gene.

Introduction

We have recently reported the enzymatic in vitro synthesis of full-length, double-stranded DNA copies of rabbit globin mRNA (Efstratiadis et al., 1976). The purpose of that work was to develop a method for gene purification and amplification which could be applied to any eucaryotic structural gene whose mRNA could be obtained. In particular, we proposed that this approach, combined with recently developed methods for molecular cloning, should make it possible to isolate a single structural gene starting with a heterogeneous mixture of mRNAs such as those found in differentiating cells. This gene could then be studied directly, or used as a hybridization probe to identify or isolate DNA sequences adjacent to the gene in chromosomal DNA.

We chose rabbit globin mRNA for these studies because it is the most thoroughly characterized eucaryotic mRNA presently available. The RNA (consisting of α- and β-chain globin mRNAs) can be obtained in high purity. Portions of the RNA have been sequenced, and the entire sequence of α- and β-globin proteins is known. It should therefore be possible to use this information in determining whether the synthetic double-stranded globin DNA is a faithful copy of globin mRNA, and in detecting sequence rearrangements during cloning and subsequent amplification in bacteria if they occur.

In this paper, we report further characterization of the double-stranded globin DNA, the construction of hybrid DNA molecules containing both globin and bacterial plasmid DNA sequences, their propagation in bacteria, and the characterization of the inserted globin DNA sequences after amplification.

Double-stranded globin DNA was prepared as described (Efstratiadis et al., 1976). Briefly, highly purified globin mRNA was reverse-transcribed by AMV reverse transcriptase using conditions known to increase the yield of full-length cDNA which contains a small hairpin sequence at the 3′ terminus (Efstratiadis et al., 1975). After alkaline hydrolysis of the mRNA, a second DNA strand (sDNA) was synthesized by incubating total cDNA template-primer with E. coli DNA polymerase I. We have shown that at this stage of the synthesis, the sDNA is covalently linked to the 3′ end of the cDNA (hairpin DNA; Efstratiadis et al., 1976). The DNA was then treated with S1 nuclease which cleaved the loop connecting the two strands and destroyed contaminating single-stranded DNA. The full-length open form of the globin DNA was then separated from the incomplete double-stranded DNA fragments by preparative polyacrylamide gel electrophoresis. The purified globin DNA was then linked to plasmid DNA using the poly(dA)·poly(dT) tailing method. The protocol for gene synthesis and amplification from mRNA is summarized in Figure 1.

Results

Physical Mapping of Synthetic Globin DNA

Synthesis and amplification of eucaryotic genes starting with purified mRNA will be useful only if the mRNA is copied into DNA without error and sequence rearrangements during the construction, cloning, and amplification of molecular hybrids. To examine the question of fidelity, we have derived a detailed restriction map of the synthetic globin DNA using restriction enzymes with known recognition sequences. We reasoned that the fidelity of globin DNA synthesis from mRNA could be evaluated by comparing the experimentally derived distances between restriction sites in the synthetic globin DNA with those between the restriction sites that could be identified in the nucleotide sequence

AUTHOR INDEX

A

Abraham, Judith A., 18
Alberts, Bruce, 31
Albrecht-Buehler, Guenter, 65, 68

B

Bar-Sagi Dafna, 69, 74
Beach, David, 25, 30
Bell, Stephen P., 50
Botchan, Michael, 113, 118
Brizuela, Leonardo, 30
Broach, James, 192
Broker, Thomas R., 112
Buchkovich, Karen J., 207
Bukhari, A.I., 11

C

Cameron, Scott, 192
Cheng, Xiaodong, 54
Chow, Louis T., 105, 112
Clarke, Jennifer, 138
Creager, Angela N.H., 209
Curran, Tom, 202

D

Dean, Caroline, 24
Del Vecchio, M., 168
Draetta, Giulio, 30

E

Efstratiadis, Argiris, 234

F

Fasano, Ottavio, 192
Feramisco, James R., 69, 74
Fink, Gerald R., 1
Franza, B. Robert, Jr., 199, 202
Friend, Stephen H., 207
Futcher, A. Bruce, 81

G

Gelinas, Richard E., 112
Garrels James I., 219, 222
Gluzman, Yakov, 128
Goldfarb, Mitchell, 175, 179, 192
Greider, Carol W., 75, 81
Grodzicker, Terri, 99, 103, 132

H

Hanahan, Douglas, 193, 197
Harley, Calvin B., 81
Harlow, Edward, 203, 207
Harshey, Rasika, 7
Haward, Samuel, 24
Herr, Winship, 133, 138
Hicks, James B., 13, 18
Hockfeld, S., 162
Horowitz, Jonathan M., 207

I

Ivy, John M., 18

J

Jones, Jonathan D.G., 24
Josephs, Steven F., 202

K

Kafatos, Fotis C., 234
Kandel, Eric, 139
Kataoka, Tohru, 192
Kee, Sim Gek, 234
Keller, Walter, 97
Kilmašauskas, Saulius, 54
Klar, Amar J.S., 18
Kostura, Matthew, 45
Kumar, Sanjay, 54

L

Lazarides, Elias, 64
Levine, Arnold J., 169
Ljungquist, E., 11
Luca, Frank, 30

M

Ma, Hong, 24
Maniatis, Tom, 227, 234
Marshak, Daniel R., 45
Martienssen, Robert, 19, 24
Mathews, Michael B., 45
McGill, Carolyn, 18, 24
McKay, Ronald, 145, 156, 157, 162

N

Nasmyth, Kim A., 18

P

Prelich, Gregory, 45
Perucho, Manuel, 179
Powers, Scott, 187, 192

Q

Quinn, W.G., 168

R

Rauscher, Frank J. III, 202
Raybuck, Margaret, 207
Roberts, Richard J., 51, 54, 112, 215, 217

Ruderman, Joan, 30
Ruley, Earl, 181, 186

S

Sambrook, Joseph F., 37, 89, 95, 97, 103, 118,
    125, 223, 226
Sharp, Phillip A., 97, 103, 226
Shimizu, Kenji, 179
Spector, David L., 55, 83, 87
Springer, Patricia, 24
Stillman, Bruce, 41, 45, 47, 50
Strathern, Jeffrey N., 18
Sugden, Bill, 226
Sundaresan, Venkatesan, 24
Szybalski, W., 12

T

Taylor, A.L., 12
Thummel, Carl S., 129, 132
Tijan, Robert, 119, 124, 132
Topp, William, 118
Tully, Tim, 163, 168

V

Volpe, Thomas, 24

W

Wallach, J.S., 168
Watson, J.D., 40
Weber, Klaus, 61, 64
Weinberg, Robert A., 207
Westendorf, Joanne, 30
Whyte, Peter, 207
Wigler, Michael, 179, 192
Wilder, E.L., 168
Williams, J., 103

Y

Yin, J. C. P., 168

Z

Zhou, H., 168
Zipser, Birgit, 151, 156

INDEX

A

Abraham, Judy, 189
Actin, 61–64
Activator protein-1 (AP-1), 171,
     199, 202
Adenovirus
    E1A inactivation of Rb, 172,
        203–207
    E1A transformation studies,
        181–183, 186
    genome mapping, 92, 100–103
    T antigen expression under
        promoters, 129–130, 132
Albrecht-Buehler, Guenter, 56–57,
     65–66
Allshire, Robin, 77
Anesthesia-resistant memory
     (ARM), 164–166
Anesthesia-sensitive memory
     (ASM), 164
AP-1. *See* Activator protein-1
Arabidopsis, 3–4, 20–21
ARM. *See* Anesthesia-resistant
     memory
ARS. *See* Autonomously replicating
     sequence
Ascher, Philip, 140
ASM. *See* Anesthesia-sensitive
     memory
Autonomously replicating sequence
     (ARS), 47–48, 50
Axel, Richard, 170, 176

B

Bacillus subtilis, 119, 121
Bacteriophage Mu, 7–9, 105

Bacteriophage T7, 37–40
Baltimore, David, 111
Bar-Sagi Dafna, 69–72, 74
Bate, Mike, 148
Beach, David, 25
Beckwith, Jon, 99
Bell, Steve, 48
Benzer, Seymour, 140,
     148–149, 158
Bernhard, Wilhelm, 83
Bevan, Mike, 20
Birchmeier, Carmen, 178
Bishop, Michael, 170
Bishop, Michael, 175
Blackburn, Elizabeth, 58, 75
Blose, Steve, 57
Bodmer, Walter, 146–147
Booher, Bob, 26, 27
Botchan, Mike, 113–114
Bowtell, David, 20
Boyer, Herb, 230
Brenner, Sydney, 140, 148
Brinkley, Bill, 58
Brinster, Ralph, 194
Broach, Jim, 188–190
Broker, Tom, 105–106,
     108–109
Bukhari, Ahmad, 7–12, 105
Burridge, Keith, 56–57, 146
Busch, Harris, 85

C

Caenorhabditis elegans, 140
Cairns, John, 32, 89–90
Cdc2, 25–30
Cdc25, 189–190

Changeux, Jean-Pierre, 140
Chen, Lan Bo, 56
Cheng, Xiadodong, 52–53
Chory, Joanne, 20
Chow, Louise, 105–106
Cooke, Howard, 77
Cone, Dick, 140
COS cell, 126
Costantini, Frank, 194
Cowan, Max, 140
Creager, Angela N. H., 209
CREB, 165–166, 168
Crick, Francis, 31–32, 142–144, 210
Curran, Tom, 200
Cyclins, 26–30

D

Darnell, Jim, 108, 110
Daughter cells, 68
Davern, Rick, 90
Davidson, Norman, 106, 110, 147
Davis, Ron, 47
Dean, Caroline, 20
Delbrück, Max, 140, 148
Delius, Hajo, 105, 109
de Thé, Guy Blaudin,
Diffley, John, 47
DNA affinity precipitation assay, 199
DNA methyltrnsferase, 51–54
Doty, Paul, 31
Downey, Kathleen, 43
Draetta, Giulio, 25–27
Drosophila, 19–20, 121, 140, 148–149,
    163–165, 188, 211, 230
Dubey, Ashok, 52
Dulbecco, Renato, 90, 111
Dunn, Ashley, 110

E

E1A, 172, 182–183, 186, 203–207
Ebbinghaus, Hermann, 164
Echols, Harrison, 211–212
Ecker, Joe, 20
Efstratiadis, Argiris, 211, 228–229, 231

Electron microscopy, 55–59, 105, 108
Escherichia coli, 7–9, 16, 89, 99, 105, 122, 173,
    177, 229–230
Ethidium bromide, 224, 226

F

Fakan, Stanislav, 83
Feramisco, Jim, 57, 69–72, 74
Fey, Georg, 41, 120, 122
Fink, Gerry, 13, 16
Fogh, Jorgen, 176
Fos, 171, 194, 202
Frank, Eric, 140
Franza, Bob, 199–200
Frederiksen, Kristen, 148
Freeling, Mike, 19
Fried, Mike, 181
Futcher, Bruce, 76–77

G

Garrels, Jim, 57, 211–212, 219–220
Gelinas, Richard, 107
Gensel, Susan, 109
Gesteland, Ray, 13, 15, 100
Gibson, Wade, 219
Gilbert, Wally, 133, 136, 144
β-Globin, gene amplification and
    characterization, 234
Gluzman, Yasha, 41–42, 119, 122, 125–127,
    134–136
Goldberg, David, 230
Goldfarb, Mitch, 175, 177
Goldman, Bob, 56, 146
Goldstein, Sam, 76
Goodenough, Ursula, 230
Goodman, Corey, 157
Graessmann, Adolf, 41, 57, 69, 120, 122
Gray, John, 22
Greene, Lloyd, 70
Greene, Ronni, 225
Greider, Carol, 58, 75–76
Grodzicker, Terri, 92, 99, 101, 119, 121,
    129–130
Gruss, Peter, 158

H

Haemophilus influenza, 223
Haemophilus parainfluenza, 211, 223
Hall, Jeff, 166
Hall, Zack, 140
Hanahan, Doug, 92, 133, 136, 173, 193
Harley, Carl, 76–78
Harlow, Ed, 133, 136, 172, 195, 203–204
Harshey, Rasika, 9
Harvey sarcoma virus, 176
Hassell, John, 110
Hastie, Nick, 77
Haward, Sam, 22
Hayflick, Leonard, 76
Heineman, Steve, 140
Heisenberg, Martin, 140
Herr, Winship, 133, 135, 199
Herskowitz, Ira, 13, 190
HhaI methyltrnsferase, 51–54
Hicks, Jim, 13–14, 16, 190
Hinnen, Albert, 13
Hirt, Bernhard, 91
Hockfield, Susan, 147, 157–158
Hogness, Dave, 211, 230
Holtzer, Howard, 61
Homeobox genes, 158
Hubbard, Ruth, 230
Hubel, David, 140, 145
Hudspeth, Jim, 140
Human papillomavirus, 109
Hunt, Tim, 26, 43

J

Jessell, Tom, 157
Jones, Jonathan, 20
Josephs, Steve, 200
Joyner, Alex, 158
Jun, 171

K

Kafatos, Fotis, 211, 228–229
Kahmann, Regina, 105
Kaiser, Dale, 230
Kamp, Dietmar, 105

Kamoroff, Lydia Villa, 230
Kandel, Eric, 130, 139, 164
Karlow, Ed, 92
Katz, Bernard, 147
Kaufmann, Berwind, 55
Kee, Sim Gek, 211
Kehoe, Jac Sue, 140
Keller, Walter, 92, 95
Kelly, Regis, 140
Kelly, Thomas, 41
Khorana, Gobin, 227
Klar, Amar, 13–14, 16
Klessig, Dan, 110
Klimasauskas, Saulius, 53
Klingenstein, John, 142
Koester, John, 141
Köhler, Georges, 146, 151, 157
Konopka, Ron, 149
Kornberg, Arthur, 12, 31
Kostura, Matt, 43
Kuffler, Steve, 140, 145
Kuhn, Thomas, 209
Kumar, Sanjay, 52–53

L

Lacy, Liz, 194, 230
Land, Harmut, 183
Lander, Art, 157
Lazarides, Elias, 56, 61, 66
Leder, Phil, 194
Lee, Wen Hwa, 205
Leventhal, Cyrus, 140
Levine, Arnie, 169
Levitt, Pat, 158–159
Lewin, Ben, 27, 29, 135
Lewis, Jim, 107, 110
Lin, Jim, 57
Littlefield, John, 90
Livingston, David, 205–206
Lobban, Peter, 230
Long-term memory (LTM), 164–166, 168
Losick, Rich, 119, 121
LTM. *See* Long-term memory

M

Ma, Hong, 21
Maniatis, Tom, 114, 211–212, 227
MAPK. *See* Mitogen-activated protein kinase
Marahrens, York, 47–48
Martienssen, Rob, 20
Martin, Gail, 158
Martin, Kelsey, 141
MAT locus, 13–18
Mathews, Mike, 43, 114, 116, 183–184
Matthei, Heinrich, 210
Maturation promoting factor (MPF), 25–30
McClintock, Barbara, 2–3, 7, 9, 19–20, 22, 100
        148, 199
McDougall, Jim, 67
McKay, Ron, 141, 145, 151, 153–154, 157
McMann, Jack, 140
Messenger RNA splicing, 108–109
Meyerowitz, Eliot, 21
Middle-term memory (MTM), 164
Milstein, Cesar, 146, 151, 157
Mitogen-activated protein kinase (MAPK), 172
Molecular Cloning: A Laboratory Manual, 210
Monoclonal antibody, leech neurone
        staining, 156
Morange, Michel, 211
Moses, Kevin, 20
Moyzis, Bob, 77
MPF. *See* Maturation promoting factor
MTM. *See* Middle-term memory
Mu transposition, 7–11
Mulder, Carel, 91
Myc, 171, 187, 195
Myersa, Phyllis, 231

N

Nasmyth, Kim, 25, 189–191
Nerve growth factor (NGF), 70
Nestin, 159–160
Neuromuscular junction (NMJ), 147
Newbold, Robert, 183
Newport, John, 26
NF-κB. *See* Nuclear factor-κB
NGF. *See* Nerve growth factor
Nichols, John, 140

Nick translation, 114
Nirenberg, Marshall, 210
Nixon, Richard, 91
NMJ. *See* Neuromuscular junction
Noller, Harry, 133, 136
Nottebaum, Fernando, 141
Nuclear factor-κB (NF-κB), 199
Numa, Shigetata, 141
Nurse, Paul, 25

O

Okazaki, Tomoko, 32
Okazaki, Tomoko, 32
Olovnikov, Alexis, 76
Oncogenes, 170–172, 179
ORC. *See* Origin recognition complex
Origin recognition complex (ORC), 33, 48
Overell, Bob, 183
Ozanne, Brad, 91

P

p16, 183
p53, 172, 183, 204
Pack, Bill, 140
Palmiter, Richard, 194
Pauling, Linus, 1
Pax genes, 158
PCNA. *See* Proliferating cell nuclear antigen
Pearson, Keir, 141
Perucho, Manuel, 176
Pettersson, Ulf, 92, 100, 111
Philipson, Lennart, 108, 110
Pollack, Robert, 55–56
Pollard, Tom, 56
Porter, Rodny, 146
Potter, David, 147
Powers, Scott, 187–188
Preat, Thomas, 165
Prelich, Greg, 42–43
Proliferating cell nuclear antigen (PCNA), 33,
        41–45
Proteomics, 219
Ptashne, Mark, 144, 227
Puvion, Edmund, 83

Q

QUEST laboratory, 211, 219–222
QUEST-2D laboratory, 57
Quinn, Chip, 163

R

Rabbitts, Terry, 231
Raff, Martin, 157
Rakic, Pasko, 158
Ras, 69–72, 74, 170–172, 177, 182–183,
    187–192
Rauscher, Frank, 200
Rb. *See* Retinoblastoma protein
REBASE, 215
Reed, Steve, 25
REF52 database, 221
Reichardt, Louis, 140, 157
Restriction endonucleases, 215–217,
    223–226, 228, 230–231
Restriction fragment length polymorphsm
    (RFLP) analysis, 100–101
Retinoblastoma protein (Rb), 183, 203–207
Reverse transcriptase, 229
RFLP analysis. *See* Restriction fragment
    length polymorphsm analysis
Rheinberger, Hans-Jörg, 210
Rich, Alex, 133–134, 136
Rio, Don, 121
RIP-Tag mouse, 195
RNase T1, 107
Roberts, James M., 106
Roberts, Rich, 19, 52, 107–109, 122, 210, 225,
    230–231
Robertson, Charles, 140
Robertson, William, 142
Rosenbaum, Joel, 56
Rous sarcoma virus (RSV), 170
RSV. *See* Rous sarcoma virus
Rubin, Gerry, 20
Ruderman, Joan, 26
Ruley, Earl, 92–100, 172, 181–182

S

S1 nuclease, 229

Saccharomyces cerevisiae, 14, 33, 47
Sambrook, Joe, 96, 99, 101, 107, 116, 181,
    195, 211, 224
Sanes, Josh, 157
Sanger, Fred, 133
Schachner, Melitta, 157
Schizosaccharomyces pombe, 4
Schubert, David, 219
Scolnick, Ed, 188
Seattle Science Foundation (SSF), 201
Sedat, John, 83
Senescence, 76–78, 81
SF. *See* Seattle Science Foundation
Sharp, Phil, 92, 100–101, 107, 109–110, 211,
    224
Shay, Jerry, 78
Shih, Chiaho, 176
Shilo, Benny, 176
Shimizu, Kenji, 177–178
Short-term memory (STM), 164
Silverstein, Saul, 170, 176
Sim, Gek Kee, 230–231
SIR genes, 14
Skarnes, Bill, 20
Small nuclear ribonucleoprotein particles
    (snRNPs), 59, 85, 87
Smart, John, 43–44
Smetana, Karel, 83
snRNPs. *See* Small nuclear ribonucleoprotein
    particles
So, Antero, 43
Spector, David, 85
Sperry, Roger, 140, 145–146, 148–149,
    151–152
Spitzer, Nick, 148
Spitzman, Lyman, 200
Springer, Patty, 22
Squire, Larry, 21, 141
SRC, 170, 175
Stent, Gunther, 140, 209
Stevens, Charles, 142
Stewart, Tim, 194
Stillman, Bruce, 33, 42
STM. *See* Short-term memory
Strathern, Jeff, 13–16, 188, 190
Sugden, Bill, 91, 122, 211

Sundaresan, Venkatesan, 19–22
supF, 177
SV40
    DNA sequence arrangement in
        transformed cells, 113–115
    enhancer elements, 133–135, 138
    genome mapping, 92
    Haemophilus parainfluenzae restriction
        enzyme studies on DNA, 223–226
    mutant replication support by
        transformed cells, 125–128
    nonsense mutations, 99, 101
    oncogene transgenic mice, 193–195, 197
    proliferating cell nuclear antigen and
        replication, 41–45
    T antigen DNA-binding site, 119–121, 124
    T antigen expression control under
        adenovirus promoters, 129–130, 132
    transcription, 95–97
Swift, Hewson, 83

T

T antigen, 119–121, 124, 126, 130, 135, 137,
    172, 182, 204
Tan, Eng, 43
Taylor, Austin, 7, 12
Telomerase, 76–78, 81
Temin, Howard, 111
Tetrahymena, 75–77
Theologis, Sakis, 20
Thomas, Charlie, 230
Thummel, Carl, 129–130
Thymidine kinase (TK), 176
Tijan, Bob, 41, 119–120, 129–130, 158
Tissieres, Alfred, 31
TK. *See* Thymidine kinase
Topp, Bill, 115–117, 183
Torpedo californica, 147
Transposon tagging, plants, 19–24
Troge, Jennifer, 135
Tully, Tim, 149, 163–164
Tummor suppressor genes, 172–174, 183
Tumor viruses, 89–93, 99–103, 170–174,
    181–183, 185
Two-dimensional gel electrophoersis,
    219–221

V

van deSande, Hans, 227
Van Essen, David, 140
Varmus, Harold, 170
Vellucci, Al, 230
Vinograd, Jerry, 90
Volpe, Tom, 22
von Ehrenstein, Gunter, 62

W

Wald, George, 230
Watson, Jim
    cancer research promotion, 172–173
    recollections, 1–2, 5, 13, 19–21, 31–32, 35,
        55–57, 66–67, 75–76, 95, 99, 105–106,
        113, 119, 139–144, 145–149, 165–166,
        190, 195, 205, 219–220, 228, 230, 232
    T7 DNA studies, 37–38, 40
    technique research promotion, 210, 212
    tumor virus research promotion, 89–91
    yeast research promotion, 16, 188
Weber, Klaus, 56, 61–62, 146
Weinberg, Bob, 170, 173, 183
Weintraub, Hal, 28–29
Wensink, Peter, 230
West, Mike, 78
Westphal, Henry, 91, 113
Wiesel, Torsten, 145
Wigler, Mike, 69, 170–171, 175–176, 178,
    187–190
Williams, Jim, 100
Wilson, E. O., 145, 149
Winocour, Ernest, 125
Wright, Woody, 78

Y

Yeast mating type, 13–18
Yin, Jerry, 149, 165–166

Z

Zain, Sayeeda, 107
Z-DNA, 133–136
Zipser, Birgit, 141, 146, 151–152, 153, 157
Zipser, David, 7, 10, 105–106, 146, 151, 154